PENGUIN BOOKS

THE PENGUIN DICTIONARY OF CURIOUS AND INTERESTING NUMBERS

David Wells was born in 1940. He had the rare distinction of being a Cambridge scholar in mathematics and failing his degree. He subsequently trained as a teacher and after working on computers and teaching machines taught mathematics and science in a primary school and mathematics in secondary schools. He is still involved with education through writing and research.

While at university he became British under-21 chess champion, and in the middle seventies was a game inventor, devising 'Guerilla' and 'Checkpoint Danger', a puzzle composer and the puzzle editor of *Games and Puzzles* magazine. From 1981 to 1983 he published *The Problem Solver*, a magazine of mathematical problems for secondary pupils. He has published several books of problems and popular mathematics, including *Can You Solve These?* and *Hidden Connections, Double Meanings* and also *Russia and England, and the Transformations of European Culture*. He has written *The Penguin Dictionary of Curious and Interesting Puzzles*, *The Penguin Dictionary of Curious and Interesting Geometry*, *The Penguin Book of Curious and Interesting Mathematics* and, also for Penguin, *You Are a Mathematician*.

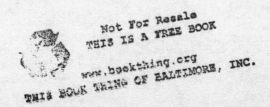

David Wells

The Penguin Dictionary of
Curious and Interesting Numbers

PENGUIN BOOKS

PENGUIN BOOKS

Published by the Penguin Group
Penguin Books Ltd, 27 Wrights Lane, London W8 5TZ, England
Penguin Books USA Inc., 375 Hudson Street, New York, New York 10014, USA
Penguin Books Australia Ltd, Ringwood, Victoria, Australia
Penguin Books Canada Ltd, 10 Alcorn Avenue, Toronto, Ontario, Canada M4V 3B2
Penguin Books (NZ) Ltd, 182–190 Wairau Road, Auckland 10, New Zealand

Penguin Books Ltd, Registered Offices: Harmondsworth, Middlesex, England

First published 1986
Reprinted with revisions 1987
Revised edition published 1997
10 9 8 7 6 5 4 3 2 1

Copyright © David Wells, 1986, 1987, 1997
All rights reserved

Filmset in Monotype Times New Roman by
Rowland Phototypesetting Ltd, Bury St Edmunds, Suffolk
Printed in England by Clays Ltd, St Ives plc

Contents

Introduction

Numbers have exercised their fascination since the dawn of civilization. Pythagoras discovered that musical harmony depended on the ratios of small whole numbers, and concluded that everything in the universe was Number. Archimedes promised the tyrant Gelon that he would calculate the number of grains of sand required to completely fill the universe, and did so.

Two thousand years later Karl Friedrich Gauss remarked that 'in arithmetic the most elegant theorems frequently arise experimentally as the result of a more or less unexpected stroke of good fortune, while their proofs lie so deeply embedded in darkness that they defeat the sharpest inquiries'.

Leopold Kronecker said that 'God himself made the integers: everything else is the work of man'.

No other branch of mathematics has been so beloved by amateurs, because nowhere else are gems so easily discovered just below the surface, aided today by pocket calculators and computers. Yet no other branch has trapped and defeated so many great mathematicians, or led them to their greatest triumphs.

This is an elementary dictionary. It presents a multitude of facts, in simple language, avoiding complicated notations and symbols. The *Glossary* explains some basic terms. Others are explained where they occur.

In contrast to a dictionary of words, it has not always been obvious where a particular property should be entered. Is the fact that $5^2 = 3^2 + 4^2$ a property of 5, or of 25? Generally speaking, if the larger number cannot be easily calculated, the entry is under the smaller number. Thus, look for properties of 144^5 under 144.

More general searches, for sums of cubes, say, may be made using the Index.

The tables at the back are for the benefit of readers who cannot wait to look for their own patterns and properties. Computers and calculators, of course, can very easily produce more extensive tables; indeed they are an indispensable aid to any modern number puzzler who is not a calculating prodigy.

One of the charms of mathematics is that good mathematics never dies. It may fade from view, but it is not demolished by later discoveries.

Aristotle's physics was primitive and rudimentary. Archimedes' mathematics still shines brilliantly. Some of these curious and interesting numbers and their properties go back to the ancient Greeks. Others were discovered only the other day, and many more are waiting to be discovered, perhaps even by you, the reader of this book. That is one of the fascinations of mathematics.

Acknowledgements

It would be impossible to credit all the sources for every property referred to. This is not a compendium of historical scholarship. I have given precedence to the discoverers, where known, and the sources, where these are unique to the best of my knowledge, of the most striking and unusual properties only.

Hundreds of books and journals have been trawled in search of curious and interesting numbers. If a particular number or property is missing, it could be that there was no room for it, even in this greatly enlarged edition, or it could be sheer ignorance on my part. Corrections and suggestions for additional entries will be welcomed.

Scores of readers have contributed suggestions for new numbers and new properties, as well as corrections to errors and infelicities, from retired professors to Samuel Vilain, aged 10. Where the reader himself, or herself, is apparently the original source of the suggestion, they have been acknowledged in the actual text. Otherwise, their contribution is acknowledged below.

A similar qualification applies to credits to authors of journal papers. Where journal references are given, if the author appears to be the originator of the conclusion, then the author is credited in the text. Otherwise, the journal reference appears without an author name.

I have included many more journal references here than in the first edition, partly because more results have come from relatively recent work, and partly to help readers who wish to follow up particular entries. However, in order to minimize the extra space involved, and to avoid distracting readers who are not concerned with original sources, I have given the minimum information sufficient to identify the reference. The journal abbreviations used in the references are as follows:

AMM   *American Mathematical Monthly*
FQ   *Fibonacci Quarterly*
JRM   *Journal of Recreational Mathematics*
MG   *Mathematical Gazette*
MOC   *Mathematics of Computation*
MM   *Mathematics Magazine*

Thus Beatty [AMM v33 159] means volume 33 of the *American Mathematical Monthly*, page 159.

Books listed in the Bibliography are referenced by author name and page number, as [Guy 221]. The exception is the *Encyclopaedia of Integer Sequences* by Sloane & Plouffe, which is referenced as [Sloane xxxxx] where xxxxx is the number of the sequence in the book.

Any sequence is almost certainly to appear in [Sloane]. Readers with access to the Internet will be pleased to know that Sloane's sequence database, which contains several thousand more sequences than the published book, is on the Internet. The address is sequence@research.att.com. The text 'lookup', followed by the terms of the sequence, with single spaces between the terms, not commas, will produce, in a few minutes, an identification of the sequence, assuming that it is in the database. For further information, see the article in *American Scientist*, Jan–Feb 1996, pp.10–14.

It should not be assumed that because a result is not given a reference to his book, that no relevant reference will be found in [Guy].

I am especially grateful for their very detailed and lengthy comments to Tony Gardiner, David Singmaster and Joseph Myers, and to Richard Guy, author of *Unsolved Problems in Number Theory*, listed in the Bibliography, and (with John Horton Conway) *The Book of Numbers*, which appeared too late to be used in this revision.

I offer my grateful thanks to all of the following, who have corrected errors and infelicities and/or suggested new entries, and my apologies to those whose suggestions could not after all be included for reasons of space in this revised edition:

Robert Baillie, Lewis Baxter, Arthur Benjamin, Miriam Berg, Emma Blount, J. Bryant, Aidan Burns, J. G. D. Carpenter, George Chambers, Robin Cooper, Frank Cousins, Michel Criton, Andrew Crompton, Lionel Crown, Michael Ede, Andy Edwards, J. Ellul, Donald Entwisle, Michele Fanelli, John Finger, Werner Goerke, C. G. Giles, Hans Glunkler, C. Gray, J. Groenewoud, S. S. Gupta, Holm Hagman, Neil Hamilton-Smith, S. J. Harber, Richard Harris, Jacques Haubrich, Chris Hawkins, W. Heybroek, Andreas Hinz, L. A. Holford-Strevens, Kenneth Hooton, Malcolm Hudson, Robert Irwin, Ken Korbin, K. Kubat, Karl Kuhn, N. T. Jones, M. E. Larsen, Jack Leet, M. Lehning, I. W. van Luttervelt, R. H. Macmillan, D. C. Maxwell, Don McDonald, Gábor Megyesi, Colin Mills, R. S. Moore, Mike Mudge, Bernard Murphy, M. J. Olney, Robert Pargeter, Richard Peto, Bernardo Recaman, Tiger Redman, D. N. Rhodes, David Roberts, J. P. Robertson, John Sharp, N. S. Srigirinath, Michal Stajsczak, Alan Stanier, Alan Stein, M. Thayer, A. J. Turner, M. Unavane, L. J. Upton,

A List of Mathematicians
in Chronological Sequence

Ahmes	c.1650 BC
Pythagoras	c.540 BC
Hippocrates	c.440 BC
Plato	c.430–c.349 BC
Hippias	c.425 BC
Theaetetus	c.417–369 BC
Archytas	c.400 BC
Xenocrates	396–314 BC
Theodorus	c.390 BC
Aristotle	384–322 BC
Menaechmus	c.350 BC
Euclid	c.300 BC
Archimedes	c.287–212 BC
Nicomedes	c.240 BC
Eratosthenes	c.230 BC
Diocles	c.180 BC
Hipparchus	c.180–c.125 BC
Heron of Alexandria	c.75
Ptolemy	c.85–c.165
Nicomachus of Gerasa	c.100
Theon of Smyrna	c.125
Diophantus	1st or 3rd century
Pappus	c.320
Iamblichus	c.325
Proclus	410–485
Tsu Ch'ung-Chi	430–501
Brahmagupta	c.628
Al-Khwarizmi	c.825
Thabit ibn Qurra	836–901
Mahavira	c.850
Bhaskara	1114–c.1185
Leonardo of Pisa, called Fibonacci	c.1170–after 1240
al-Banna, Ibn	1256–1321

Chu Shih-chieh	early 14th century, *c.*1303
Pacioli, Fra Luca	*c.*1445–1517
Leonardo da Vinci	1452–1519
Dürer, Albrecht	1471–1528
Stifel, Michael	1486/7–1567
Tartaglia, Niccolò	*c.*1500–1557
Cardano, Girolamo (also known as Cardan)	1501–1576
Recorde, Robert	*c.*1510–1558
Ferrari, Ludovico	1522–1565
Viète, François	1540–1603
Ceulen, Ludolph van	1540–1610
Stevin, Simon	1548–1620
Napier, John	1550–1617
Cataldi, Pietro Antonio	1552–1626
Briggs, Henry	1561–1630
Kepler, Johannes	1571–1630
Oughtred, William	*c.*1574–1660
Bachet, Claude-Gaspar, de Meziriac	1581–1638
Mersenne, Marin	1588–1648
Girard, Albert	*c.*1590–*c.*1633
Desargues, Girard	1591–1661
Descartes, René	1596–1650
Fermat, Pierre de	1601–1665
Brouncker, Lord William	*c.*1620–1684
Pascal, Blaise	1623–1662
Huygens, Christian	1628–1695
Newton, Isaac	1642–1727
Leibniz, Gottfried Wilhelm	1646–1716
Bernoulli, Johann	1667–1748
Machin, John	1680–1751
Bernoulli, Nikolaus	1687–1759
Goldbach, Christian	1690–1764
Stirling, James	1692–1770
Euler, Leonhard	1707–1783
Buffon, Count Georges	1707–1788
Lambert, Johann	1728–1777
Lagrange, Joseph Louis	1736–1813
Wilson, John	1741–1793
Wessel, Caspar	1745–1818
Laplace, Pierre Simon de	1749–1827
Legendre, Adrien Marie	1752–1833

Nieuwland, Pieter	1764–1794
Ruffini, Paolo	1765–1822
Argand, Jean Robert	1768–1822
Gauss, Karl Friedrich	1777–1855
Brianchon, Charles	c.1783–1864
Binet, Jacques-Philippe-Marie	1786–1856
Möbius, August Ferdinand	1790–1868
Babbage, Charles	1792–1871
Lamé, Gabriel	1795–1870
Steiner, Jakob	1796–1863
de Morgan, Augustus	1806–1871
Liouville, Joseph	1809–1882
Shanks, William	1812–1882
Catalan, Eugène Charles	1814–1894
Hermite, Charles	1822–1901
Riemann, Bernard	1826–1866
Venn, John	1834–1923
Lucas, Edouard	1842–1891
Cantor, George	1845–1918
Lindemann, Ferdinand	1852–1939
Hilbert, David	1862–1943
Lehmer, D. N.	1867–1938
Hardy, G. H.	1877–1947
Ramanujan, Srinivasa	1887–1920

Glossary

The term 'number' means 'whole number' unless otherwise qualified. I have used it frequently in place of 'integer', for brevity and for the sake of variety.

Similarly, 'square', 'cube' and so on, unless otherwise qualified, mean 'perfect square' (the square of an integer), 'perfect cube' and so on.

BIQUADRATE   An old-fashioned term for a fourth power, a number multiplied by itself three times. $10 \times 10 \times 10 \times 10 = 10{,}000$, and so 10,000 is a biquadrate.

COMPOSITE   A composite number is an integer that has at least one proper factor. $14 = 2 \times 7$, as well as 14×1, is composite. 13, which only equals 13×1, is not; it is prime.

CUBE   A number that is equal to another number multiplied by itself twice. $216 = 6 \times 6 \times 6$, and therefore 216 is a cube. *See* PERFECT SQUARE.

DIGIT   The digits of 142857 are the numbers 1, 4, 2, 8, 5 and 7. Occasionally a number is written with initial zeros, for example 07923. When this is done, the initial zero is ignored when the number of digits is counted, so 07923 counts as a 4-digit number.

DIVISOR   An integer that divides another integer exactly. The divisors of 10 are 10, 5, 2 and 1. DIVISOR and FACTOR are synonyms in this dictionary.

PROPER DIVISOR (or PROPER FACTOR)   A divisor of a number which is not the number itself, or 1. The proper divisors of 10 are 5 and 2, only.

EGYPTIAN FRACTION   A unit fraction, a fraction with unit numerator, so called because they were used almost exclusively in Egyptian papyri.

FACTOR   *See* DIVISOR.

FACTORIAL   Factorial n, or n factorial, usually written $n!$ and often pronounced 'n bang!', means the product $1 \times 2 \times 3 \times 4 \times 5 \times \ldots \times (n-1) \times n$. For example, 6 factorial $= 6! = 1 \times 2 \times 3 \times 4 \times 5 \times 6 = 720$.

HYPOTENUSE   The Greek term for the longest side of a right-angled triangle, the one opposite the right-angle. In the well-known $3-4-5$ right-angled triangle, the side of length 5 is the hypotenuse.

INTEGER   A whole number.

IRRATIONAL   Any real number that is not rational, and therefore any number that *cannot* be written as a decimal that either terminates or repeats. The numbers $\pi = 3.14159\,265\ldots$; $e = 2.71828\,18\ldots$ and $\sqrt{2} = 1.41421\ldots$ are all irrational.

MULTIPLE   A multiple of an integer is any other integer that the first integer divides without remainder. If P is a *multiple* of Q, then Q is a *factor* of P. Any integer has infinitely many multiples, because it can be multiplied by any other integer.

OF THE FORM   This phrase, like REPRESENTED AS, is used to indicate that a number is equal to an expression of a certain type. For example, all primes, except 2 and 3, are of the form $6n \pm 1$, meaning that every prime is either 1 more or less than a multiple of 6. 17 is of the form $6n \pm 1$, because it is in fact equal to $6 \times 3 - 1$.

PERFECT SQUARE   An integer that is the square of another integer. In other words, its square root is also an integer. $25 = 5^2$ and $144 = 12^2$ are perfect squares. In this book it will usually be taken for granted that SQUARE means PERFECT SQUARE, and similarly CUBE means PERFECT CUBE and so on.

PERMUTATION   A permutation of a sequence of objects is just a rearrangement of them. EBDCA is a permutation of ABCDE.

CYCLIC PERMUTATION   A permutation is cyclic if it merely takes some objects from one end and transfers them, without changing their order, to the other end. CDEAB is a cyclic permutation of ABCDE.

POWER   In this book, power will be a general term for squares, cubes and higher powers.

PRIME   A prime number is an integer greater than 1 with no factors apart from itself and 1. 17 is prime because the only integers dividing it without remainder are 17 and 1.

PRIMORIAL   Primorial (p), often denoted by $p\#$, is only defined if p is a prime, when it is equal to the product of all the primes up to and including p. So, for example, $11\# = 2 \times 3 \times 5 \times 7 \times 11$.

PRODUCT   The product of several numbers is the result of multiplying them all together. The product of the first five prime numbers equals $2 \times 3 \times 5 \times 7 \times 11 = 2310$.

RATIONAL   Any number that is either an integer or a fraction (the ratio of two integers). All rational numbers can be written as decimals that either terminate or repeat. For example, $1/7 = 0.14285\,71428\,57\ldots$ and $1/8 = 0.125$. *See* IRRATIONAL.

RECIPROCAL   Only reciprocals of integers are referred to in this dictionary. The reciprocal of an integer n is the fraction $1/n$.

REPRESENTED AS   This phrase, like OF THE FORM, is used to state that a number is equal to an expression of a certain type. For example, 25 can be represented as the sum of two squares, because $25 = 16 + 9$ and 16 and 9 are both squares. *See* OF THE FORM.

REPUNIT   A number all of whose digits are units.

ROOT   The square root of a number n, written $\sqrt{n}$, is the number that must be multiplied by itself to produce n. Since $7 \times 7 = 49$, $\sqrt{49} = 7$. The cube root of a number n, written $\sqrt[3]{n}$, is the number that must be multiplied by itself twice to produce n. Since $5 \times 5 \times 5 = 125$, $\sqrt[3]{125} = 5$. Fourth roots, and higher roots (fifth roots, sixth roots and so on) are defined in the same way. For example, since $2 \times 2 \times 2 \times 2 \times 2 = 32$, the fifth root of 32, written $\sqrt[5]{32}$, $= 2$.

SQUARE   The square of a number is the number multiplied by itself. Thus 12 squared, written 12^2, $= 12 \times 12 = 144$.

TRANSCENDENTAL NUMBER   A real number that does *not* satisfy any algebraic equation with integral coefficients, such as $x^3 - 5x + 11 = 0$. All transcendental numbers are irrational and can be written, in theory, as non-terminating, non-repeating decimals. Most irrational numbers are transcendental.

UNIT FRACTION   The reciprocal of an integer. 1/13 and 1/28 are unit fractions. 2/3 is not.

$\phi(n)$, pronounced 'phi [fie] n' is the number of integers less than n, and having no common factor with n. So $\phi(13) = 12$, because 13 is prime, and $\phi(6) = 2$, because the only numbers less than 6 and prime to it are 1 and 5.

$d(n)$ is the number of factors of n, including unity and n itself.

$\sigma(n)$, pronounced 'sigma n' is the sum of all the factors of n, including unity and n itself. So $\sigma(6) = 1 + 2 + 3 + 6 = 12$.

All three functions are listed in Table 8.

Bibliography

Books

The following books all contain considerable material on numbers, and are all readily available from libraries. The items marked with * are at a higher level.

Not readily available, and at a considerably higher level than this book, is another dictionary, *Les Nombres remarquables*, by François Le Lionnais, Hermann, Paris 1983.

A superbly detailed guide to all aspects of recreational mathematics is *A Bibliography of Recreational Mathematics* by William L. Schaaf, published in the USA by the National Council of Teachers of Mathematics in four paperback volumes. My edition has no ISBN, but their address is 1906 Association Drive, Reston, Virginia 22091, USA.

BALL, W. W. R., and COXETER, H. S. M., *Mathematical Recreations and Essays*, University of Toronto Press, 1974

BEILER, ALBERT H., *Recreations in the Theory of Numbers*, Dover, New York, 1964

*DICKSON, L. E., *A History of the Theory of Numbers*, 3 vols., Chelsea Publishing Co., New York, 1952

DUDENEY, H. E., *Amusements in Mathematics*, Nelson, London, 1951 (Other books of puzzles by Dudeney also contain some numerical material.)

GARDNER, MARTIN, *Mathematical Puzzles and Diversions*, Penguin, Harmondsworth, 1965

——, *More Mathematical Puzzles and Diversions*, Penguin, Harmondsworth, 1966

——, *Martin Gardner's Sixth Book of Games from Scientific American*, W. H. Freeman, San Francisco, 1971

——, *Mathematical Carnival*, Penguin, Harmondsworth, 1975

——, *Mathematical Circus*, Penguin, Harmondsworth, 1979

——, *Further Mathematical Diversions*, Penguin, Harmondsworth, 1981

——, *New Mathematical Diversions from Scientific American*, University of Chicago Press, Chicago, 1984

(Readers are warned that Gardner's books often change their titles in crossing the Atlantic.)

*GUY, RICHARD K., *Unsolved Problems in Number Theory*, Springer-Verlag, New York, 2nd edition, 1994

HUNTER, J. A. H., and MADACHY, JOSEPH S., *Mathematical Diversions*, D. von Nostrand Co., New York, 1963

KORDEMSKY, BORIS A., *The Moscow Puzzles*, Puzzles, Harmondsworth, 1976

KRAITCHIK, MAURICE, *Mathematical Recreations*, George Allen & Unwin, London, 1960

MADACHY, JOSEPH S., *Mathematics on Vacation*, Charles Scribner, New York, 1966

RIBENBOIM, PAULO, *The Book of Prime Number Records*, Springer-Verlag, New York, 2nd edition, 1989

SLOANE, N. J. A. and PLOUFFE, SIMON, *Encyclopaedia of Integer Sequences*, Academic Press, New York, 1995

Magazines and Journals

For readers with a mathematical background scores of professional journals have occasional material of recreational interest. The six journals listed in the Acknowledgements are especially promising.

Libraries may also have sets of two magazines, now ceased: *Recreational Mathematics* magazine, and *Scripta Mathematica*, which is not as obscure as its title suggests.

Mathematics teachers' journals are also a fertile source of ideas, and many schools and colleges produce their own small magazines. For example, Cambridge University students publish *Eureka* and *Quarch*, whose purpose is to promote discussion of famous, interesting and unsolved problems of a recreational nature.

The Dictionary

−1 and *i*

negative and complex numbers

At the age of 4, Paul Erdös remarked to his mother, 'If you subtract 250 from 100, you get 150 below zero.' Erdös could already multiply 3- and 4-digit numbers together in his head, but no one had taught him about negative numbers. 'It was an independent discovery,' he recalls happily. [Tierney, 'Paul Erdös is in town. His brain is open', *Science*, Oct. 1984]

Erdös grew up to be a great mathematician, but a surprising number of schoolchildren without his extraordinary talent will answer the question, 'How might this sequence continue: 8 7 6 5 4 3 2 1 0 . . . ?' by suggesting, '1 less than nothing!' or 'minus 1, minus 2 . . . !'

Children in our society are floating in numbers. Whole numbers, fractions, decimals, approximations, estimations, record-breaking large numbers, minusculely small numbers. The *Guinness Book of Records* is a twentieth-century Book of Numbers, including the largest number in this Dictionary.

A mere handful of centuries ago numbers were smaller, fewer and simpler. It was seldom necessary to count beyond a few thousand. The Greek word *myriad*, which suggests a vast horde, was actually a mere 10,000, a fair size for an entire Greek army, but to us a poor attendance at a Saturday football match.

Fractions often stopped at one-twelfth. Merchants avoided finer divisions by dividing each measure into smaller measures, and the small measures into yet smaller, without going as far as Augustus de Morgan's fleas: 'Great fleas have little fleas upon their backs to bite 'em/And little fleas have lesser fleas, and so ad infinitum.'

The very conception of numbers proceeding to infinity, in any direction, appeared only in the imaginations of theologians and the greatest astronomers and mathematicians, such as Archimedes, who exhausted a circle with indefinitely many polygons and counted the grains of sand required to fill the universe.

To almost everyone else, numbers started at 1 and continued upwards in strictly one direction only, no further than ingenious systems of finger arithmetic, or the clerk's counting board, allowed.

(Zero, a strange and brilliant Indian invention, is not used for counting anyway. The Greeks had no conception of a zero number.)

These numbers were solid and substantial. To Pythagoras and his followers a number was always a number of things. To arrange a number such as 16 in a square pattern of dots was their idea of advanced and abstract mathematics.

To merchants also, numbers counted things.

To the later Greeks, numbers were still lengths of lines, areas of plane figures, or volumes of solids. What does a sphere with a volume −10 look like?

How could they make sense of numbers less than zero? Early mathematicians did sometimes bump into negative numbers, in the dark as it were. They tried to avoid them, or pretended that they were not there, that they were an illusion.

Diophantus was a pioneer in number theory who still thought in strongly geometrical language. He solved many equations that to us have one negative and one positive root. He accepted the positive and rejected the negative. He 'knew' it was there, but it made no sense. If an equation had no positive root, he rejected the equation. $x + 10 = 5$ was not a proper equation.

Perhaps it was a misfortune for a number-theorist to be born Greek. The Indians did not think of mathematics as geometry. Hindu mathematicians first recognized negative roots, and the two square roots of a positive number, and multiplied positive and negative numbers together, though they were suspicious also. Bhaskara commented on the negative root of a quadratic equation, 'The second value is in this case not to be taken, for it is inadequate; people do not approve of negative roots.'

On the other hand, the Chinese had already discovered negative numbers for counting purposes. By the twelfth century they were freely using red counting rods for positive quantities and black rods for negative, the exact opposite of our bank statements before computerization. They did not, however, recognize negative roots of equations.

As any schoolteacher will recognize, a chasm separates the simple act of counting backwards from the idea that negative numbers can be operated on in the same manner as positive numbers (with a couple of provisos). How many generations of schoolchildren have never progressed further than the magic incantation, 'Two minuses make a plus!'

Craftsmen do not need negative numbers to measure backwards along a line. They turn their ruler round, or hold the ruler firmly and walk round the length they are measuring.

Merchants and bank clerks may easily juggle credits and debits without any conception that they are subtracting one negative number from another. Their intentions are honourably practical and concrete.

In fact, they made a practical contribution to the notation of mathematics. Our familiar plus and minus signs were first used in fifteenth-century German warehouses to show when a container was over or under the standard weight.

Number-theorists had a different problem. They met negative numbers stark naked, in the abstract. The number that when added to 10 makes 5 is just a number – or is it a fake number?

[4]

Renaissance mathematicians were as distrustful as Diophantus or Bhaskara. Michael Stifel talked of numbers that are 'absurd' or 'fictitious below zero', which are obtained by subtracting ordinary numbers from zero. Descartes and Pascal agreed. Yet, in the early Renaissance, one of the most difficult known problems was the solutions of equations, which often cried out for negative solutions. A few mathematicians accepted them, and even took a giant step further. Cardan was one.

The solutions to quadratic equations had been known since the Greeks, though Renaissance mathematicians continued to recognize three different types, illustrated by $x^2 = 5x + 6$, $x^2 + 5x = 6$, and $x^2 + 6 = 5x$. No negative coefficients!

The cubic equation was much harder. Cardan, in his book *The Great Art*, still presented the cubic in more than a dozen different varieties, and solved them, using an idea he took from Tartaglia. Yet he recognized negative numbers and even approached their square roots. [Cardan, *Ars Magna*, 1545]

The very first square root of negative number on record, $\sqrt{81 - 144}$, is in the *Stereometrica* of Hero of Alexandria. Another, $\sqrt{1849 - 2016}$ was met by Diophantus as a possible root of a quadratic equation. They did not take them seriously. Neither did fifteenth-century European mathematicians.

Cardan proposed the problem: Divide 10 into 2 parts such that the product is 40. He first said it was obviously impossible, but then solved it anyway, correctly giving the two solutions, $5 + \sqrt{-15}$ and $5 - \sqrt{-15}$. He concluded by telling the reader that 'These quantities are "truly sophisticated" and that to continue working with them would be "as subtle as it would be useless".'

The square roots of negative numbers! If negative numbers were false, absurd or fictitious, it is hardly to be wondered at that their square roots were described as 'imaginary'. Even today, the theory of complex numbers is one of several hurdles that are recognized as separating 'elementary' from 'advanced' mathematics.

Paul Erdös's most famous proof is of the Prime Number theorem, which says that if $\pi(x)$ is the number of primes not exceeding x, then as x tends to infinity,

$$\frac{\pi(x) \log x}{x}$$

tends to 1.

It was originally proved in 1896 using complex analysis. Here, 'complex' does not mean complicated, though it was, but using complex numbers. Erdös in 1949 published a proof that avoided complex numbers entirely. Such a proof is called 'elementary'. Here 'elementary' does not mean easy, merely that complex numbers are *not* used!

John Wallis accepted negative numbers but wrote of complex numbers, 'These Imaginary Quantities (as they are commonly called) arising from the Supposed Root of a Negative Square (when they happen) are reputed to imply that the Case proposed is Impossible.' Wallis sounds (if I may say so) when talking of complex numbers (when he does) much like Bhaskara on numbers less than zero.

Mathematicians had reasons to be suspicious. Negative numbers, quintessentially -1, do possess properties that positive numbers lack.

A friend of Pascal, Antoine Arnauld, argued that if negative numbers exist, then $-1/1$ must equal $1/-1$, which seems to assert that the ratio of a smaller to a larger quantity is equal to the ratio of the same larger quantity to the same smaller. Most educated adults today would reject this idea after a moment's thought. No wonder this paradox was discussed at length.

Complex numbers are even more fiendish. Is $\sqrt{-1}$ less than or greater than, say, 10? Neither, as Euler realized. The very idea of greater than or less than breaks down, and has to be reconstructed in a new form, a form incidentally that will also resolve Arnauld's paradox.

Fortunately, negative and complex numbers work, just as the calculator's red and black rods, or the warehouseman's $+$ and $-$ signs work. Mathematicians were forced to accept negative and imaginary numbers, long before they had solved the conundrums that they posed.

Euler boldly used $\sqrt{-1}$ in infinite series, and published his exquisite formula $e^{i\pi} = -1$. He also introduced the letter i to stand for $\sqrt{-1}$.

Wessel, Argand and Gauss independently discovered around 1800 that complex numbers could be represented on a graph. When Gauss introduced the term 'complex number' and expressed complex numbers as number pairs, their modern conception was almost complete.

[F. Cajori, *A History of Mathematical Notations*, 2 vols., Open Court, 1977 (reprint); Augustus de Morgan, *A Budget of Paradoxes*, 1872]

0
Zero

A mysterious number, which started life as a space on a counting board, turned into a written notice that a space was present, that is to say that something was absent, then confused medieval mathematicians who could not decide whether it was really a number or not, and achieved its highest status in modern abstract mathematics in which numbers are defined anyway only by their properties, and the properties of zero are at least as clear, and rather more substantial, than those of many other numbers.

The Babylonians in the second century BC used a system for mathematical and astronomical work in which the value of a numeral depended on its position. Two small wedges indicated that a place within a number was unoccupied, so distinguishing 207 from 27. (270 was distinguished from 27 by context alone.) Whether this Babylonian system was transmitted to neighbouring cultures is not known.

Our system, in which the 0 is an extra numeral, originated in India. It was used from the second century BC to denote an empty place and as a numeral in a book by Bakhshali published in the third century. The Sanskrit name for zero was *sunya*, meaning empty or blank, as it does today in some Indian languages. Translated by the Arabs as *sifr*, with the same meaning, it became the European name for nought, via the Latin *zephirum*, in different ways in different countries: *zero*, *cifre*, *cifra*, and the English words *zero* and *cipher*.

In AD 773 there appeared at the court of Caliph Al-Mansur in Baghdad an Indian who brought writings on astronomy by Brahmagupta. This was read by Al-Khwarizmi, the great Arab mathematician, whose name gave us the word 'algorithm' for an arithmetical process and more recently for a wider class of processes such as computers use, and who wrote a textbook of arithmetic in which he explained the new Indian numerals, published in AD 820.

At the other end of the Muslim world, in Spain at the beginning of the twelfth century, it was translated by Robert of Chester. This translation is the earliest known description of Indian numerals to the West. There are several records of Arabic, that is, Indian, numerals being taught over the next century and a half. About 1240 they were even taught in a long and not very good poem. Yet they spread very slowly indeed, for two reasons.

The Arabic system did not just add a useful zero to the old Roman numerals; learners had to master the Arabic numerals 1 to 9 as well, and the zero numeral was a puzzle in itself. Was zero a number? Was it a digit? If it stands for nothing, then surely it is nothing? But as every school pupil knows, if you add a harmless zero to the end of a number, you multiply it by 10! Our ten digits were often presented as the digits 1 to 9, plus the cipher, the zero: 'And there are nine figures that have value . . . and one more figure outside of them which is called null, 0, which has no value in itself but increases the value of others.'

The twelfth-century Salem Monastery manuscript had sounded a Platonic note: 'Every number arises from One, and this in turn from the Zero. In this lies a great and sacred mystery' though Plato started with One and knew nothing of any zero.

[7]

Merchants and bookkeepers had another reason to hesitate. To avoid tampering with written records, important amounts of money were written in full, in which case Indian numerals have no advantage, useful though they were for actual calculation.

A decisive step was taken by the first great mathematician of the Christian West, Leonardo of Pisa, called Fibonacci, who also features in this dictionary as the discoverer of the Fibonacci sequence. Leonardo gives details of his life in his most famous book, the *Liber Abaci*. Leonardo's father was the chief magistrate of the Pisan trading colony at Bugia in Algeria. Leonardo spent several years in Africa, studying under a Muslim teacher. He also travelled widely to Greece, Egypt and the Middle East.

No doubt many merchants before Leonardo had noticed that the merchants they traded with used a very different system of numerals. Leonardo compared the systems he met, and concluded that the Indian system he had learned in Africa was by far the best. In 1202, and in a revised edition in 1228, he published his Book of Computation, the *Liber Abaci*, a compendium of almost all the mathematics then known. In it he described the Indian system. Having learned of it as a merchant's son, he described its use in commercial arithmetic, in calculating proportions and mixtures, and in exchanging currency.

The final practical triumph of zero and its Indian numerals came with the spread of the printed book, and the rise of the merchant class.

Textbooks of arithmetic were among the most popular of the early printed books. They taught the merchant's children the skills with numbers that were becoming more and more essential at the same time as they gave the final push to counters and the counting board, and established the new numerals.

We so easily take zero for granted as a number that it is surprising to consider that the Greeks had no conception of nothing, or emptiness, as a number, and doubly curious that this did not stop them, or many other cultures, from creating mathematics. Even when the Greeks treated limits and very small quantities, they had no conception of a quantity 'tending to zero'. It was sufficient that the quantity was less than another quantity, or might be made as small as desired.

Familiarity with zero did not exhaust its interest for mathematicians, who anyway had some problems in handling this extraordinary number. Brahmagupta stated that 'positive or negative divided by cipher is a fraction with that for denominator'. This was called 'the quantity with zero as denominator'. Mahavira wrote in his *Compendium of Calculations*: 'A number multiplied by zero is zero and that number remains unchanged which is divided by, added to or diminished by zero.' Did he think of division by zero as repeated subtraction, which had no effect?

The fact that zero added to or subtracted from a number left the number unchanged was a mystery directly comparable to the Pythagoreans' refusal to accept 1 as a number, since it did not increase other numbers by multiplication.

Both these facts are part of the abstract definition of a field, of which ordinary numbers are an example. A field must contain a 'multiplicative identity', usually labelled 1 with the property that if g is any other element in the field, then $1 \times g = g \times 1 = g$, and an 'additive identity', usually labelled 0, with the properties that for any g, $0 + g = g + 0 = g$, and division by 0 is forbidden.

Like unity, 0 proves exceptional in other ways. It is an old puzzle to decide what 0^0 means. Since a^0 is always 1, when a is not zero, surely by continuity it should also equal 1 when a is zero? Not so! 0^a is always 0, when a is not zero, so by the same argument from continuity, 0^0 should equal 0. [Karl Menninger, *Number Words and Number Symbols*, MIT Press, 1969]

The low status of zero in some circumstances is a great advantage to the lucky mathematician. When Lander and Parkin were looking for 5 5th powers whose sum was also a 5th power, one of their solutions included the number 0^5. This solution immediately qualified, because powers of 0 do not count for obvious reasons, as a sum of 4 5th powers equal to a 5th power, and destroyed a conjecture of Euler. (*See 144.*)

0·02010 30407 11 ...
Equal to 199/9899. It displays the start of the Lucas sequence, 2, 1, 3, 4, 7, 11 ... The same sequence appears more spread-out in the fraction 1999/998,999, and so on.

0·11000 10000 00000 00000 00010 00000 00000 00000 0 ...
Liouville's number, equal to $10^{-1!} + 10^{-2!} + 10^{-3!} + 10^{-4!} + \ldots$

Liouville proved in 1844 that transcendental numbers actually do exist by constructing several, of which this is the simplest. Cantor later proved that almost all numbers are transcendental.

0·12345 67891 01112 13141 51617 18192 02122 ...
The digits of this number are the natural numbers in sequence. Like Liouville's number, and π and e, it is transcendental.

It is also normal, that is, whether expressed in base 10, or any other base, each digit occurs in the long run with equal frequency. It is not known whether π and e are normal.

Tests of the square roots of the integers 2 to 15 (4, 9 excluded) in bases 2, 4, 8 and 16, suggest that they are also normal. [Beyler, Metropolis and Neergaard, MOC v24]

0·20787 95763 50761 90854 6955 . . .
The value of i^i or $e^{-\pi/2}$ (where $i = \sqrt{-1}$). These two expressions are equal by Euler's relationship, $e^{i\pi} = -1$.

16/64
When Denis the Dunce reduces this fraction by cancelling the 6s, he gets the right answer, 1/4. There are just 3 similar patterns with numbers less than 100:

$$19/95 = 1/5 \quad 26/65 = 2/5 \quad 49/98 = 4/8$$

These are all examples of longer patterns. Thus $16,666/66,664 = 1/4$ also.

There are many variations on this theme:

$$3544/7531 = 344/731 \quad 143,185/17,018,560 = 1435/170,560$$

$$\frac{37^3 + 13^3}{37^3 + 24^3} = \frac{37 + 13}{37 + 24} \quad \frac{3^4 + 25^4 + 38^4}{7^4 + 20^4 + 39^4} = \frac{3 + 25 + 38}{7 + 20 + 39}$$

[Moessner, *Scripta Mathematica* v19 and v20]

0·30102 99956 63981 . . .
The logarithm of 2 to base 10. To calculate the number of digits in a power of 2, multiply the index by log 2 and take the next highest integer.

Thus, the 127th Mersenne number, $2^{127} - 1$ has 39 digits because $127 \times 0·30103 = 38·23$.

0·31830 98861 83790 67153 77675 26745 02872 40689 19291 480
π^{-1}

1/3
$1/3 = (1 + 3)/(5 + 7) = (1 + 3 + 5)/(7 + 9 + 11) = (1 + 3 + 5 + 7)/(9 + 11 + 13 + 15) . . .$ [Galileo, 1615]

0·36787 94411 71442 32159 55237 70161 46086 74458 11131 031
e^{-1}. As a sum of Egyptian fractions it is approximately:

$$1/3 + 1/29 + 1/15,786 + 1/513,429,610$$

As the number of letters and envelopes in the problem of the mis-addressed letters increases (*see 44, Subfactorial*), the probability that every letter will be placed in the wrong envelope rapidly approaches this limiting value.

The same problem may be simulated by well shuffling two packs of cards, and turning up pairs of cards, one from each pack. The probability that there will be no match among the 52 pairs is approximately e^{-1}.

0·43429 44819 03251 82765 11289 18916 60508 22943 97005 803 . . .
The logarithm of e to base 10.

0·5
1/2

There are twelve ways in which the digits 1 to 9 can be used to write a fraction equal to 1/2.

6729/13,458 has the smallest numerator and denominator, 9327/18,654 the largest. The same puzzle can be solved for other fractions.

$$1/7 = 2637/18,459$$

and the same fraction with both numbers doubled, 5274/36,918. Similarly,

$$4/5 = 9876/12,345$$

[Friedman, *Scripta Mathematica* v8]

The sum $\zeta(s) = 1 + \dfrac{1}{2^s} + \dfrac{1}{3^s} + \dfrac{1}{4^s} + \dfrac{1}{5^s} + \ldots$

can also be written as an infinite product,

$$\zeta(s) = \frac{2^s}{2^s - 1} \times \frac{3^s}{3^s - 1} \times \frac{5^s}{5^s - 1} \times \frac{7^s}{7^s - 1} \times \frac{11^s}{11^s - 1} \times \ldots$$

in which the numerators are powers of the primes. Because of this relationship many problems about the distribution of prime numbers depend on the behaviour of this function.

Riemann conjectured that, considered as a complex function with complex roots, its roots all had real part equal to 1/2. So important is this possibility that many mathematical proofs have been published that assume that Riemann's hypothesis is true.

This profound conjecture is generally considered to be the outstanding problem in mathematics today. It is known that the first $1\frac{1}{2}$ billion roots are of the conjectured form. However, many phenomena of this type are known in which trends for small numbers are misleading.

It was announced in December 1984 that the Japanese mathematician Matzumoto, working in Paris, had finally proved it, but his proof was flawed. Riemann's hypothesis remains unproved.

0·57721 56649 01532 86060 65120 90082 40243 1 . . .

γ, Euler's constant, sometimes called Mascheroni's constant, calculated by Euler to 16 places and also named *gamma* by him in 1781.

It is the limit as *n* tends to infinity of $1 + 1/2 + 1/3 + 1/4 + 1/5 + \dots + 1/n - \log n$.

It is not even known whether γ is irrational, let alone whether it is transcendental, though it is known that if it is a rational fraction *a/b*, then *b* is greater than $10^{10,000}$. [Brent, MOC v31]

0·60792 7101 . . .

$$\frac{6}{\pi^2} = \left(\frac{1}{1^2} + \frac{1}{2^2} + \frac{1}{3^2} + \frac{1}{4^2} + \frac{1}{5^2} + \dots \right)^{-1} = \zeta(2)^{-1}$$

It is the probability that if 2 numbers are chosen at random, they will have no common factor, and also the probability that one number chosen at random is not divisible by a square.

2/3

The uniquely unrepresentative 'Egyptian' fraction, since the Egyptians used only unit fractions, with this one exception. All other fractional quantities were expressed as sums of unit fractions.

From the Rhind papyrus: Divide 7 loaves among 10 men – Answer: $2/3 + 1/30$. Because they multiplied by repeated doubling, then adding, they used tables of double unit fractions. In the Rhind papyrus is a table going up to double 1/101.

$$2/7 = 1/4 + 1/28$$
$$2/11 = 1/6 + 1/66$$
$$2/97 = 1/56 + 1/679 + 1/776$$

Egyptian fractions are a fertile source of problems. For example, Erdös and Sierpinski have conjectured, respectively, that $4/n$ and $5/n$ are each expressible for all *n* as the sum of 3 unit fractions. [Guy]

0·69314 71805 59945 30941 72321 21458 17656 80755 00134 360
$\log 2$ (to base *e*) = $1 - 1/2 + 1/3 - 1/4 + 1/5 \dots$

0·7404 . . .

$$\frac{\pi}{\sqrt{18}}$$

How closely can identical spheres be packed together? The obvious way is to arrange one layer on a plane so that each sphere touches 6 others, and then arrange adjacent layers, so that each sphere touches 3 others in each layer (12 in all) and so on. However, no mathematician has been able to prove this 'obvious' fact.

If that were the closest packing, the density would be this number.

'Many mathematicians believe, and all physicists know, that the density cannot exceed $\dfrac{\pi}{\sqrt{18}}$.' [Rogers]

0·78539 81 . . .

$\pi/4$. The sum of Leibniz's series: $1/1 - 1/3 + 1/5 - 1/7 + 1/9 - 1/11 + \ldots$

0·83190 7 . . .

$1/\zeta(3)$, where $\zeta(3) = 1/1^3 + 1/2^3 + 1/3^3 + 1/4^3 + \ldots$

It is the probability that if 3 integers are chosen at random, no common factor will divide them all.

0·9068 . . .

$$\frac{\pi}{2\sqrt{3}}$$

Identical circles packed together in a plane in a hexagonal array, so that each touches 6 others, cover this proportion of the plane.

101,010,101/110,010,011

This fraction (in base 10, or any other base greater than 1) remains unchanged when the same even number of additional units are added in the middle of the numerator and denominator: for example, it is equal to $1,010,111,110,101/1,100,111,110,011$. The value is $9091/9901$.
[Sierpinski, *250 Problems in Elementary Number Theory*, no. 208]

1
Unity

The Greeks did not consider 1, or unity, to be a number at all. It was the monad, the indivisible unit from which all other numbers arose. According to Euclid a number is an aggregate composed of units. Not unreasonably, they did not consider 1 to be an aggregate of itself.

As late as 1537, the German Kobel wrote in his book on computation, 'Wherefrom thou understandest that 1 is no number, but it is a generatrix, beginning, and foundation for all other numbers.'

The special significance of 1 is apparent in our language. The words 'one', 'an' and 'a' (a shortened form of 'an') are etymologically the same.

So are the words 'unit', 'unity', 'union', 'unique' and 'universal', which all come from the Latin for one. It is no coincidence that these words are all exceptionally important in modern mathematics.

The Greeks considered that 1 was both odd and even, because when added to an even number it produced odd, and when added to an odd number it produced even. This reasoning is completely spurious, because any odd number has the same property. They were right, however, to notice that 1 is the only integer that produces more by addition than by multiplication, since multiplication by 1 does not change a number. In contrast, every other integer produces more by multiplication than by addition.

It is because multiplication by 1 does not change a number that 1 hardly ever appears as a coefficient in expressions such as $x^2 + x + 4$. It is pointless to write x as $1x$, unless we wish to emphasize some pattern.

On the other hand, 1 is of vital significance when summing infinite series. The series,

$$1 + x + x^2 + x^3 + x^4 + x^5 + x^6 + \ldots$$

has no sum if x is greater than 1, because each term is then greater than the previous term. If $x = 1$, then the series becomes $1 + 1 + 1 + 1 + 1 + 1 \ldots$ and still has no sum. But when x lies between 1 and -1, then the sum of as many terms as we choose to add approaches as closely as we wish to $1/(1 - x)$, without ever exceeding that number, and the infinite series has a finite sum.

What did the Greeks do about fractions? Surely they recognized that the indivisible unit, 1, could be divided into 2 parts, or 3 parts, or 59 parts? Not at all! They took the view that the original unit remained the same, while the result of the division, say $1/59$, was taken as a new unit. Indeed, we still talk of a fraction whose numerator is 1 as a unit fraction.

This interpretation fits the usage of merchants and craftsmen throughout the world. How much easier it is to consider 2 centimetres, rather than 0·02 metres, though they are mathematically the same! Psychologically, it is much simpler to invent new units of measure for small, and large, quantities, and completely avoid using very small or very large numbers.

1 appears in its modern disguise as the generatrix, the foundation of other numbers, in so many infinite sequences. It is, of course, the first square number, but it is also the first perfect cube, and the first 4th power, the first 5th power . . . the first of any power.

It is also the first triangular number, the first pentagonal number . . . the first Fibonacci number and the first Catalan number!

Into how many pieces can a circular pancake be cut with n straight cuts? It is natural to start with the 1 piece, the whole pancake, which remains after zero cuts.

In how many ways can n objects be arranged in order? Modern mathematicians naturally start with 1 object, which can be 'arranged' in just 1 way. The Greeks would undoubtedly have argued, very plausibly, that the sequence should start with 2 objects, which can be arranged in order in 2 ways. They would have claimed that 1 object cannot be arranged in any order at all.

1 is especially important because of its lack of factors. This suggests that it should be counted as a prime number, because it fits the definition, 'A prime number is divisible by no number except itself and 1', but once again 1 is usually considered to be an exception.

A conventional reason depends on an important and favourite theorem, that any number can be written as the product of prime factors in only one way, apart from different ways of ordering the factors. Thus $12 = 2 \times 2 \times 3$ and no other product of prime numbers equals 12.

This theorem would have to be adjusted if 1 were a prime, because then 12 would also equal $1 \times 2 \times 2 \times 3$, and $1 \times 1 \times 2 \times 2 \times 3$ and so on. Untidy! So 1 is dismissed from the list of primes.

Euler had a different reason for rejecting 1. He observed that the sum of the divisors of a prime number, p, is always $p + 1$, the prime p itself and the number 1. The exception, of course, to this rule turns out to be 1. The simplest way to dispose of this exceptional case is to deny that 1 is prime.

Because 1 is so small, as it were, and has no factors apart from itself, it does not feature in many of the properties in this dictionary. To write 1 as the sum of two squares, it is necessary to write $1 = 1^2 + 0^2$ which is trivial. In the same way, 1 can be written as the sum of 3 squares, or even of 5 cubes, which is even more boring.

Similarly, 1 is the smallest number that is simultaneously triangular and pentagonal. Also boring! Indeed, 1 might be considered to be the first number that is both boring and interesting. Yet it does appear in this dictionary in a small but essential way. Precisely because it has no factors, it is never obvious whether expressions such as $2^5 - 1$, the 5th Mersenne number, or $2^{2^3} + 1$, the 3rd Fermat number, will have any factors.

When Euclid wanted to show that the number of primes is unlimited, he considered three primes, by way of example. Call them, A, B and C. Multiply them together, and add 1: is $ABC + 1$ prime? If so there is a prime larger than any of A, B or C. If $ABC + 1$ is not prime, then it has a prime factor, which cannot be any of the primes A, B or C. So there is at least one more prime ... Euclid's argument would not have worked if he had considered $ABC + 2$, or $ABC + 3$. Only 1 will guarantee his argument.

Our number line, familiar to children in school, extends at least from 0 to infinity, and the gaps between the whole numbers are filled by infinities of fractions, irrational numbers, and even more transcendental numbers. The Greeks' idea of number was simpler and inadequate for the purposes of modern mathematicians. Yet one great mathematician saw the whole numbers, starting with 1, as the only real numbers. 'God made the integers,' claimed the nineteenth-century mathematician Kronecker. 'All the rest are the work of man.'

1 is not the first number in this dictionary, but in its own way it is the foundation on which all the other entries are based.

[Menninger, *Number Words and Number Symbols*, MIT Press, 1969]

1·06066 0 . . .
$$\frac{3\sqrt{2}}{4}$$

Prince Rupert proposed the problem of finding the largest cube that may be passed through a given cube, that is to say the size of the largest square tunnel through a cube.

Pieter Nieuwland first found the solution. In theory, making no allowance for physical constraints such as friction, a cube of side 1·06066 0 . . . may be passed through a cube of side 1. The axis of the tunnel is not parallel to a diagonal of the cube, but the edges of the original cube are divided in rational proportions, 1:3 and 3:13.

[Schrek, 'Prince Rupert's Problem', *Scripta Mathematica* v16]

1·08232 3 . . .
$$\frac{\pi^4}{90}$$

The limit of the sum $1/1^4 + 1/2^4 + 1/3^4 + 1/4^4 + \ldots$

1·20205 6 . . .
The limit of the sum $1/1^3 + 1/2^3 + 1/4^3 + \ldots$

It is relatively easy to sum the series $1/r^n$ when n is even. Euler calculated all the values from 2 to 26. The sums are all rational multiples of π^n.

It is far harder to calculate the sums for odd n. It is known that 1·202 . . . is irrational, but not whether it is transcendental.

1·25992 10498 94873 16476 . . .
$\sqrt[3]{2}$ (cube root of 2)

The duplication of the cube
The three famous problems of antiquity were the duplication of the cube, the trisection of the angle and the squaring of the circle. Ideally, the Greeks would have preferred to solve each of them using only an unmarked straight edge and a pair of compasses. The legend was told that the Athenians sent a deputation to the oracle at Delos to inquire how they might save themselves from a plague that was ravaging the city. They were instructed to double the size of the altar of Apollo.

This altar was cubical in shape, so they built a new altar twice as large in each direction. The resulting altar, being eight times the volume of the original, failed to appease the gods and the plague was unabated. To find a cube whose volume is double that of another, is equivalent to finding the cube root of 2.

The Greeks interpreted this requirement geometrically. Hippocrates showed that it was equivalent to the problem of finding 2 mean proportionals between 2 lines of length x and $2x$. In other words, to find the line segments of lengths p and q such that $x/p = p/q = q/2x$. This is impossible with ruler and compasses.

The Greeks, however, were not limited to lines and circles, and in searching for solutions they created some of the finest achievements of Greek mathematics. Archytas of Tarentum solved the problem by finding the intersection of 3 surfaces of revolution, a cone, a cylinder and a torus whose inner diameter was zero.

Menaechmus is supposed to have discovered the conic sections, the parabola, ellipse and hyperbola, while attempting to solve this problem. He solved it by finding the intersections of 2 parabolas, or alternatively by the intersection of a parabola and a hyperbola.

Two Greeks, Nicomedes and Diocles, invented curves specifically to solve the problem, called the conchoid and the cissoid respectively.

1·41421 35623 73095 04880 16887 24209 69807 85697 . . .
Root 2
The square root of 2, and length of the diagonal of a unit square.

Pythagoras or one of his school first discovered that the ratio of the diagonal of a square to its side is not a ratio of integers, that is, it is irrational. This discovery had a profound effect on the Pythagoreans, who had supposed that every phenomenon could be explained in terms of the integers.

Theodorus, who taught mathematics to Plato, subsequently proved that the square roots of the numbers from 3 up to 17 are irrational, apart from the perfect squares 4, 9 and 16. He apparently stopped at 17, for no obvious reason, but clearly did not have a general proof that every integer is either a perfect square or its square root is irrational.

A sequence of best possible approximations to root 2 is 1/1 3/2 7/5 17/12 41/29 99/70 239/169 577/408 . . .

7/5 was a Pythagorean approximation. The Babylonians used 17/12 as a rough approximation to root 2, and $1 + 24/60 + 51/60^2 + 10/60^3$ ($= 1.41421\ 55 \ldots$) as a more accurate approximation.

The fractions in this sequence are the best possible approximations for a given size of denominator. They are related by the simple rule that, if a/b is one term, the next is $(a + 2b)/(a + b)$. (Theon of Smyrna in the second century knew that if a/b is an approximation, then $(a + 2b)/(a + b)$ is a better one.)

They have many other properties. For example, every other fraction has an odd numerator and denominator. Split the numerator into the sum of two consecutive numbers:

$$\frac{41}{29} = \frac{20 + 21}{29}$$

Then, $20^2 + 21^2 = 29^2$.

They also provide solutions to Pell's equation: $x^2 - 2y^2 = \pm 1$.

$$7^2 - 2 \times 5^2 = -1$$
$$17^2 - 2 \times 12^2 = +1$$
$$41^2 - 2 \times 29^2 = -1$$

and so on.

It can also be represented as a sum of Egyptian fractions, $1 + 1/3 + 1/13 + 1/253 + 1/218,201 + \ldots$ [Sloane 2962]

In binary, $\sqrt{2} = 1.01101\ 01000\ 00 \ldots$

Roland Sprague describes a very beautiful property. Write down the multiples of root 2, ignoring the fractional parts, and underneath the numbers missing from the first sequence:

1	2	4	5	7	8	9	11	12 …
3	6	10	13	17	20	23	27	30 …

The differences between the upper and lower numbers is $2n$ in the nth place. [Sprague, *Recreations in Mathematics*, Blackie, 1963]

1·44466 7861 . . .

$e^{1/e}$

The solution to Steiner's problem: for what value of x is $x^{1/x}$ a maximum? [Dorrie, *100 Great Problems of Elementary Mathematics*, Dover, 1965]

Euler proved that the function $x^{x^{x^{x^{x^{x^{\cdots}}}}}}$, where the height of the tower of exponents tends to infinity, had a limit if x is between e^{-e} = 0·06598 8 . . . and this upper limit, $e^{1/e}$.

1·61803 39887 49894 84820 45868 34365 63811 77203 09179 80576 . . .
The Divine Proportion
The Divine Proportion or Golden Ratio, equal to

$$\frac{\sqrt{5}+1}{2}.$$

In the pentagram, which the Pythagoreans regarded as a symbol of health, the ratio AB to BC is the Golden Ratio. So is the ratio AC to AB, and similar ratios in the next figure.

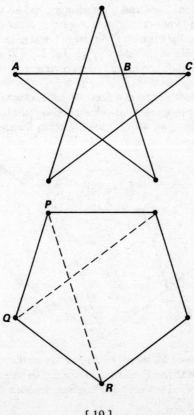

Euclid in his *Elements* calls this division 'in the extreme and mean ratio' and used it to construct first a regular pentagon, then the two most complex Platonic solids, the dodecahedron, which has 12 pentagonal faces, and the icosahedron, which is its dual. The mystical significance of these beautiful polyhedra to the Greeks was naturally transferred to the Golden Ratio.

There is some evidence that the ratio was important to the Egyptians. The Rhind papyrus refers to a 'sacred ratio' and the ratio in the Great Pyramid at Gizeh of an altitude of a face to half the side of the base is almost exactly 1·618. The Greeks probably used it in architecture but no documentary proof remains. There is no doubt that it was exploited by Renaissance artists, who knew it as the Divine Proportion.

Fra Luca Pacioli published in 1509 *De Divina Proportione*, illustrated with drawings of the Platonic solids made by his friend Leonardo da Vinci. Leonardo was probably the first to refer to it as the 'sectio aurea', the Golden Section. The Greeks, surprisingly, had no short term for it. Pacioli presented 13 of its remarkable properties, concluding that 'for the sake of salvation, the list must end (here)', because 13 was the number present at the table at the Last Supper. Fra Luca also reduced the 8 standard operations of arithmetic to 7 in reverence to the 7 gifts of the Holy Ghost.

'The Ninth Most Excellent Effect' is that 2 diagonals of a regular pentagon, as in the figure above, divide each other in the Divine Proportion. Tie an ordinary knot in a strip of paper, carefully flatten it, and the same figure appears.

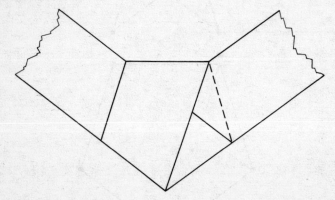

Kepler, who based his theory of the heavens on the 5 Platonic solids, enthused over the Divine Proportion, declaring 'Geometry has two great treasures, one is the Theorem of Pythagoras, the other the division of a line into extreme and mean ratio; the first we may compare to a measure of gold, the second we may name a precious jewel.'

It has been claimed that Renaissance artists regularly used the Golden Section in dividing the surface of a painting into pleasing proportions, just as architects naturally used it to analyse the proportions of a building. The first Italian edition of *De Architectura* by Vitruvius uses the Golden Ratio to analyse the elevation of Milan Cathedral.

The psychologist Gustav Fechner revived this aesthetic aspect of the Golden Ratio in his attempts to set aesthetics on an experimental basis.

He endlessly measured the dimensions of pictures, cards, books, snuff-boxes, writing paper and windows, among other things, in an attempt to develop experimental aesthetics 'from below'. He concluded that the preferred rectangle had its sides in the Golden Ratio.

Le Corbusier, the architect, followed this belief in its efficacy in designing The Modular. He constructed 2 series in parallel, one of powers of the Golden Ratio, and the other of double these powers. A fellow architect detected the double influence of the Renaissance and the Gothic spirit in it, and correspondents rushed to support Le Corbusier's claims for its harmonizing properties.

Mathematicians now either call the Golden Ratio τ, first letter of the Greek *tome*, to cut, or they use the Greek letter ϕ, following the example of Mark Barr, an American mathematician, who named it after Phidias, the Greek sculptor.

If the greater part of the line is of length ϕ and the lesser part 1, then

$$\frac{\phi + 1}{\phi} = \frac{\phi}{1}$$

which may also be written as

$$\phi^2 = \phi + 1, \text{ or as } \frac{1}{\phi} = \phi - 1$$

In other words, it is squared by adding unity, $(1 \cdot 618 \ldots)^2 = 2 \cdot 618 \ldots$ and its reciprocal is found by subtracting unity,

$$\frac{1}{\phi} = \frac{\sqrt{5} - 1}{2}$$

(Occasionally its reciprocal is called the Golden Ratio, which can be slightly confusing.)

If a rectangle is drawn whose sides are in the Golden Ratio, it may be divided into a square and another, similar, rectangle. This process may be repeated *ad infinitum*.

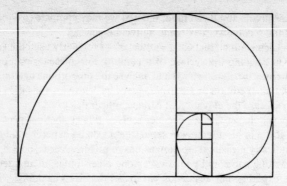

It is possible to draw an equiangular spiral through successive vertices of the sequence of rectangles. The diagram shows an excellent approximation to this spiral, a sequence of quarter circles. The spiral tends towards the point where the diagonals of all the Golden Rectangles meet.

The Golden Ratio itself is intimately related to the Fibonacci sequence. Like ϕ^2, the higher powers of ϕ can all be expressed very simply in terms of ϕ:

$$\phi^2 \qquad \phi^3 \qquad \phi^4 \qquad \phi^5 \qquad \phi^6$$
$$\phi + 1 \quad 2\phi + 1 \quad 3\phi + 2 \quad 5\phi + 3 \quad 8\phi + 5$$

Each power is the sum of the 2 previous powers, and the coefficients of ϕ form the Fibonacci sequence over again, as do the integer parts of the powers.

ϕ has many other properties. It is equal to the simplest continued fraction:

$$1 + \cfrac{1}{1 + \cfrac{1}{1 + \cfrac{1}{1 + \ldots}}}$$

which is also the slowest of all continued fractions to converge to its limit. The successive convergents are $1/1 \quad 2/1 \quad 3/2 \quad 5/3 \ldots$ the numerators and denominators following the Fibonacci sequence. Two easy to remember approximations are $377/233$ and $233/144$. Coincidentally, $355/113$ is an excellent approximation to π.

Beatty [AMM v33 159] explains a more obscure but equally beautiful property: calculate the multiples of ϕ and ϕ^2 by the whole numbers, 0, 1, 2, 3, 4, 5 ... rejecting the fractional parts. The result is a sequence of pairs: (0, 0), (1, 2), (3, 5), (4, 7), (6, 10), (8, 13), (9, 15) ...

This sequence has the triple property that the differences between the numbers in each successive pair increase by 1; the smaller number in each pair is the smallest whole number that has not yet appeared in the sequence, and the sequence includes every whole number exactly once. As a final flourish, these pairs of numbers are all the winning combinations in Wythoff's game.

[Huntley, *The Divine Proportion*, NCTM, 1970]

1·64493 4066 ...

$$\frac{\pi}{6} = \frac{1}{1^2} + \frac{1}{2^2} + \frac{1}{3^2} + \frac{1}{4^2} + \dots$$

$$= \frac{2}{1} \times \frac{2}{3} \times \frac{3}{2} \times \frac{3}{4} \times \frac{5}{4} \times \frac{5}{6} \times \frac{7}{6} \times \frac{7}{8} \times \frac{11}{10} \times \frac{11}{12} \times \frac{13}{12} \times \frac{13}{14} \dots$$

1·73205 08075 68877 29352 74463 41505 87236 69428 ...

The square root of 3, the 2nd number, after root 2, to be proved irrational, by Theodorus.

Archimedes gave the approximations, $1351/780 < \sqrt{3} < 265/153$ (or $26 - 1/52 < 15\sqrt{3} < 26 - 1/51$).

These satisfy the equations $1351^2 - 3 \times 780^2 = 1$ and $265^2 - 3 \times 153^2 = -2$, which are consistent with the view that Archimedes had some understanding of Pell equations.

1·77245 38509 05516 02729 81674 83341 14518 27975 ...

$$\sqrt{\pi} = \Gamma(\tfrac{1}{2})$$

The factorial function, $n!$, which is defined for all positive integers and by convention for 0, can be defined by means of an integral for non-integral values of n. This function is denoted by $\Gamma(n + 1)$. $\Gamma(\tfrac{1}{2}) = \sqrt{\pi}$.

1·83928 67552 1416 ...

The Tribonacci constant, the only real root of $x^3 - x^2 - x - 1 = 0$, which is related to Tribonacci sequences (in which $u_n = u_{n-1} + u_{n-2} + u_{n-3}$) as the Golden Ratio is related to the Fibonacci sequence and its generalizations. This ratio also appears when a snub cube is inscribed in an octahedron or a cube, by analogy once again with the appearance of the Golden Ratio when an icosahedron is inscribed in an octahedron.

[John Sharp, 1997]

1·90216 054 ...

The approximate value of Brun's constant, equal to the sum $1/3 + 1/5 + 1/5 + 1/7 + 1/11 + 1/13 + 1/17 + 1/19 + 1/29 + 1/31 + \dots$

where the denominators are the twin primes. [Ribenboim 201] It is sometimes calculated without the repetition of 1/5.

(It has also been calculated starting $1 + 1/3 + 1/3 + 1/5 + 1/5 + 1/7 + \ldots$ leading one mathematician briefly and optimistically to conjecture that its sum was π.)

It is not known if the number of prime pairs is infinite. However, this sum is known to converge, in contrast to the sum of the reciprocals of the primes, which diverges. Its value is exceedingly hard to calculate. The best estimate is $1 \cdot 90195 \pm 10^{-5}$.

2

The number 2 has been exceptional from the earliest times, in many aspects of human life, not just mathematically. It is distinguished in many languages, for example in original Indo-European, Egyptian, Arabic, Hebrew, Sanskrit and Greek, by the presence of dual cases for nouns, used when referring to 2 of the object, rather than 1 or many. A few languages also had trial and quaternal forms.

The word two, when used as an adjective, was often inflected, as were occasionally the words three and four. Modern languages reflect the significance of 2 in words such as dual, duel, couple, pair, twin and double.

The early Greeks were uncertain as to whether 2 was a number at all, observing that it has, as it were, a beginning and an end but no middle. More mathematically, they pointed out that $2 + 2 = 2 \times 2$, or indeed that any number multiplied by 2 is equal to the same number added to itself. Since they expected multiplication to do more than mere addition, they considered 2 an exceptional case. Whether 2 qualified as a proper number or not, it was considered to be female, as were all even numbers, in contrast to odd numbers, which were male.

Division into 2 parts, dichotomy, is more significant psychologically and more frequent in practice than any other classification.

The commonest symmetry is bilateral, 2-sided about a single axis, and is of order 2. Our bodies are bilaterally symmetrical, and we naturally distinguish right from left, up from down, in front from behind. Night is separated from day, there are 2 sexes, the seasons are expressed in pairs of pairs, summer and winter separated by spring and autumn, and comparisons are most commonly dichotomous, such as stronger or weaker than, better or worse than, youth versus age and so on.

2 and division into 2 parts is just as significant in mathematics. 2 is the first even number, all numbers being divided into odd and even. The basic operations of addition, subtraction, multiplication and division are binary operations, performed in the first instance on 2 numbers. By subtraction from zero, every positive number is associated with a unique negative

number, and 0 divides all numbers into positive and negative. Similarly division into 1 associates each number with its reciprocal.

2 is the first prime and the only even prime.

2 is a factor of 10, the base of the usual number system. Therefore a number is divisible by 2 if its unit digit is, and by 2^n if 2^n divides the number formed by its last n digits.

Powers of 2 appear more frequently in mathematics than those of any other number.

A positive integer is the sum of two or more consecutive integers if and only if it is *not* a power of 2.

The first deficient number. All powers of a prime are deficient, but powers of 2 are only just so.

Euler asserted what Descartes had supposed, that in all simple polyhedra, for example the cube and the square pyramid, the number of vertices plus the number of faces exceeds the number of edges by 2.

Fermat's last theorem states that the equation $x^n + y^n = z^n$ has solutions in integers only when $n = 2$. The solutions are then sides of a right-angled Pythagorean triangle.

Fermat's equation being exceedingly difficult to solve, several mathematicians have noticed in an idle moment that $n^x + n^y = n^z$ is much easier. Its only solutions in integers are when $n = 2$, and $2^1 + 2^1 = 2^2$.

Also, the only solution of $3^x + 4^y = 5^z$ in integers is $x = y = z = 2$. The equation $5^x + 12^y = 13^z$ has the same unique solution. [*Acta Arithmetica* v63 351]

Goldbach conjectured that every even number greater than 2 is the sum of 2 prime numbers.

The binary system

The English imperial system of measures used to contain a long sequence of measures, some of which are still in use, in which each measure was double the previous one. Presumably they were very useful in practice, though it is unlikely that most merchants had any idea how many gills were contained in a tun:

1 tun = 2 pipes = 4 hogsheads = 8 barrels = 16 kilderkins = 32 firkins or bushels = 64 demi-bushels = 128 pecks = 256 gallons = 512 pottles = 1024 quarts = 2048 pints = 4096 chopins = 8192 gills.

The numbers appearing in this list are just powers of 2, from $2^0 = 1$ up to $2^{13} = 8192$. These measures could very easily have been expressed in binary notation, or base 2.

Every number can be expressed in a unique way as the sum of powers of 2. Thus: $87 = 64 + 16 + 4 + 2 + 1$, which can be written briefly as $87 = 1,010,111$. Each unit indicates a power of 2 that must be included and each zero a power that must be left out, as in this chart for 1,010,111:

64	32	16	8	4	2	1
yes	no	yes	no	yes	yes	yes
1	0	1	0	1	1	1

The binary system was invented in Europe by Leibniz, although it is referred to in a Chinese book which supposedly dates from about 3000 BC. Leibniz associated the 1 with God and the 0 with nothingness, and found a mystical significance in the fact that all numbers could thus be created out of unity and nothingness. Without accepting his mathematical theology we can appreciate that there is immense elegance and simplicity in the binary system.

As long ago as 1725 Basile Bouchon invented a device that used a roll of perforated paper to control the warp threads on a mechanical loom. Any position on a piece of paper can be thought of as either punched or not-punched. The same idea was used in the pianola, a mechanical piano popular in Victorian homes, which was also controlled by rolls of paper.

The looms were soon changed to control by punched cards, which were also used in Charles Babbage's Analytical Engine, a forerunner of the modern digital computer, which relied on punched cards until the arrival of magnetic tapes and discs. Binary notation is especially useful in computers because they are most simply built out of components that have two states: either they are on or off, full or empty, occupied or unoccupied. The same principle makes binary notation ideal for coding messages to be sent along a wire. The 1 and 0 are represented by the current being switched on and off.

Long before mechanical computers were invented, the Egyptians multiplied by doubling, as many times as necessary, and adding the results. For example, to multiply by 6 it is sufficient to double twice, and add the 2 answers together. Within living memory, Russian peasants used a more sophisticated version of the same idea, which was once used in many parts of Europe. To multiply 27 by 35, write the numbers at the top of 2 columns: choose one column and halve the number again and again, ignoring any remainders, until 1 is reached. Now double the other number as many times:

27	35
13	70
6	~~140~~
3	280
1	560
	945

Cross out the numbers in this second column that are opposite an even number in the first. The sum of the remaining numbers is the answer.

One of the simplest and most basic facts about a number is its parity, whether it is odd or even, that is, whether it is divided by 2 without remainder. All primes are odd, except 2. All known perfect numbers are even.

The sum of this series:

$$\frac{1}{1^n} + \frac{1}{2^n} + \frac{1}{3^n} + \frac{1}{4^n} + \dots$$

is far easier to calculate if n is even than if it is odd.

The simplest kind of symmetry is twofold, as when ink is dropped on to a sheet of paper, and the paper is folded once and pressed down to produce a symmetrical blot.

Parity appears in well-known puzzles such as Sam Lloyd's 'Fifteen' puzzle. Every possible position of the tiles can be classified as either odd or even. If the position you are attempting to reach is of opposite parity from your starting position, you may as well give up and go home. It is impossible to reach.

2·09455 1 . . .
The real solution to the equation $x^3 - 2x - 5 = 0$.

This equation was solved by Wallis to illustrate Newton's method for the numerical solution of equations. It has since served as a test for many subsequent methods of approximation, and its real root is now known to 4000 digits.

[Gruenberger, 'Computer Recreations', *Scientific American*, April 1984]

2·30258 50929 94045 68401 79914 54684 36420 7601 . . .
The natural logarithm of 10.

2·50662 8 . . .
$\sqrt{2\pi}$

The constant factor in Stirling's asymptotic formula for $n!$ and therefore the limit as n tends to infinity of

$$\frac{n!e^n}{n^n\sqrt{n}}$$

2·61803 3 . . .
The square of ϕ, the Golden Ratio, and the only possible number such that $\sqrt{n} = n - 1$.

2.66514 4 . . .
$2^{\sqrt{2}}$

The 7th of Hilbert's famous 23 problems proposed at the 1900 Mathematical Congress was to prove the irrationality and transcendence of certain numbers. Hilbert gave as examples $2^{\sqrt{2}}$ and e^{π}. Later in his life he expressed the view that this problem was more difficult than the problems of Riemann's hypothesis or Fermat's Last Theorem. Nevertheless, e^{π} was proved transcendental in 1929 and $2^{\sqrt{2}}$ in 1930, illustrating the extreme difficulty of anticipating the future progress of mathematics and the real difficulty of any problem – until after it has been solved.

2·71828 18284 59045 23536 02874 71352 66249 77572 47093 69995 . . .
e, the base of natural logarithms, also called Napierian logarithms, though Napier had no conception of base and certainly did not use e.

It was named 'e' by Euler, who proved that it is the limit as x tends to infinity of $(1 + 1/x)^x$. It also equals the limit as n tends to infinity

of $\dfrac{n}{\sqrt[n]{n!}}$.

Newton had shown in 1665 that $e^x = 1 + x + x^2/2! + x^3/3! + \ldots$ from which $e = 1 + 1 + 1/2! + 1/3! + 1/4! + \ldots$ a series which is suitable for calculation because its terms decrease so rapidly.

By chance, the first few decimal places of e are exceptionally easy to remember, by the pattern 2·7 1828 1828 45 90 45 . . .

The best approximation to e using numbers below 1000 is also easy to recall: $878/323 = 2·71826 \ldots$

Like π, e is irrational, as Euler proved in 1737.

Hermite proved that e is also transcendental in 1873.

e features in Euler's beautiful relationship, $e^{i\pi} = -1$ and, more generally, e is related to the trigonometrical functions by $e^{i\theta} = \cos\theta + i\sin\theta$. It possesses the remarkable property that the rate of change of e^x at $x = t$, is e^t, from which follows its importance in the differential and integral calculus, and its unique role as the base of natural logarithms.

If you select real numbers between 0 and 1 until the total exceeds 1, the expected number of selections is e. [MG v74 167]

There are many approximate relations between e and π, most of them rather crude. This one is surprisingly accurate:

$$\sqrt[9]{10 \times e^8} = 3·14159\ 828 \ldots \text{ [Michele Fanelli]}$$

3
The first odd number according to the Greeks, who did not consider unity to be a number. To the Pythagoreans, the first number because, unlike 1

and 2, it possesses a beginning, and middle and an end. They also considered 3, and all odd numbers, to be male, in contrast to even numbers, which were female. The first number, according to Proclus, because it is increased more by multiplication than by addition, meaning that 3×3 is greater than $3 + 3$.

Division or classification into 3 parts is exceptionally common. In many languages, the positive, comparative and superlative are differentiated. In English the sequence once–twice–thrice goes no further.

There were trinities of gods in Greece, Egypt and Babylon. In Christianity, God is a trinity.

In Greek mythology there were 3 Fates, 3 Furies, 3 Graces, 3 times 3 Muses, and Paris had to choose between 3 goddesses.

Oaths are traditionally repeated 3 times. In the New Testament, Peter denies Christ three times. The Bellman in 'The Hunting of the Snark' says, more prosaically, 'What I tell you three times is true!'

The world is traditionally divided into 3 parts, the underworld, the earth, and the heavens.

The natural world is 3-dimensional, Einstein's 4th dimension of time being unsymmetrically related to the 3 dimensions of length. In 3 dimensions, at most 3 mutually perpendicular lines can be drawn.

The Greeks considered lengths, the squares of lengths, which were represented by areas, and the cubes of lengths, represented by solids. Higher powers were rejected as unnatural. Numbers with 3 factors were sometimes considered as solid, just as a number with 2 factors was interpreted by a plane figure, such as a square or some shape of rectangle, or by one of the polygonal figures. (A commentator on Plato describes even numbers as isosceles, because they can be divided into equal parts, and odd numbers as scalene.)

They also associated 3 with the triangle, which has 3 vertices and 3 edges, and was the commonest figure in their geometry and ours.

The trisection of the angle was one of the three famous problems of antiquity, the others being the squaring of the circle, and the duplication of the cube. The problem is, or was, to trisect an arbitrary angle, using only a ruler, meaning an unmarked straight edge, and a pair of compasses. Like the duplication of the cube, it depends, in modern language, on the solution of a cubic equation.

Descartes showed that this can be accomplished as the intersection of a parabola and a circle, but unfortunately the required points on the parabola cannot be constructed by ruler and compasses.

It can however be solved by the use of special curves. Pappus used a hyperbola, and Hippias invented the quadratrix, and Diocles the cissoid, which can be used to divide an angle in any proportion. The conchoid invented by Nicomedes will trisect the angle and duplicate the cube.

Euler proved that in any triangle, the centroid lies on the line joining the circumcentre to the point of intersection of the altitudes, and divides it in the ratio 1:2.

A circle can be drawn through any 3 points not on a straight line.

There are just 3 tesselations of the plane with regular polygons, using equilateral triangles, squares, or hexagons as in a honeycomb.

3 is the second triangular number, after the inevitable 1. Gauss proved that every integer is the sum of at most 3 triangular numbers. The 18th entry in his diary, dated 10 July 1796, when he was only 19 years of age, reads EYPHKA! num = $\Delta + \Delta + \Delta$.

All numbers that are not of the form $4^n(8m + 7)$ are the sum of 3 squares.

3 divides 1 less than any power of 10. Consequently a number is divisible by 3 if and only if the sum of its digits is divisible by 3.

3 is the second prime, and the first odd prime, the first prime of the form $4n + 3$, and the first Mersenne prime, since $3 = 2^2 - 1$.

It is the first Fermat prime, $3 = 2^{2^0} + 1$.

All sufficiently large odd numbers are the sum of at most three primes. [Vinogradov, 1937]

It is the first member of a prime pair, 3 and 5, the next few pairs being (5, 7), (11, 13), (17, 19), (29, 31), (41, 43) . . . It is not known if the number of prime pairs is infinite.

It is the first member of an arithmetical progression of 3 primes, 3–5–7.

$3 = 1! + 2!$

The cubic case of Fermat's Last Theorem, $x^3 + y^3 = z^3$ has no solution in integers, proved by Euler.

The smallest magic square is of order 3.

$\phi(3) = \phi(6)$. Together with $\phi(5) = \phi(8)$ the only solutions less than a million to the identity $\phi(n) = \phi(n + 3)$. [Guy 90]

At any party with 6 or more people, there are either 3 who are mutual acquaintances or 3 who are mutual strangers.

The volume of the smallest tetrahedron with integer edges and integer volume is 3. There are 2 possible sets of values for the edges: 32, 33, 35, 44, 70, 76; and 21, 32, 47, 56, 58, 76. [Dove & Sumner, MM v65 104]

The value of the infinite nested root, $\sqrt{1 + \sqrt{2 + \sqrt{3 + \sqrt{4}}}} + \ldots$ [Ramanujan: MM v59 23]

The product of 3 consecutive integers is never a perfect power. [Newman, *A Problem Seminar*, problem 32] π

π, the most famous and most remarkable of all numbers, is the ratio of the circumference of a circle to its diameter, and the area of a unit circle, and also the ratio of the circumference of any convex curve of constant diameter to its diameter (Barbier's theorem).

π is the only irrational and transcendental number that occurs naturally, if only as a rough approximation, in every society where circles are measured.

In the Old Testament, I Kings 7:23 implies that π is equal to 3. The Babylonians about 2000 BC supposed that π was either 3 or $3\frac{1}{8}$. The Egyptian scribe Ahmes, in the Rhind papyrus (1500 BC), stated that the area of a circle equals that of the square of 8/9 of its diameter, which makes π equal to (16/9) squared or 3·16049 . . .

Such crude values were adequate for primitive craftsmen or engineers. To the Greeks however, who were the first 'pure' mathematicians, π had a deeper significance. They were fascinated by the problem of 'squaring the circle', one of the 'three famous problems of antiquity', that is, of finding by a geometrical construction, using ruler and compasses only, a square whose area was exactly, not approximately, equal to a given circle.

Archimedes, by calculating the areas of regular polygons with 96 sides, determined that π lay between $3\frac{10}{71} = 3\cdot14085$ . . . and $3\frac{10}{70} = 3\cdot142857$ . . . Archimedes also found more accurate approximations to the value of π. This last value is $3\frac{1}{7}$ or 22/7, known to generations of schoolchildren. It is also the best approximation to π, using the ratio of two numbers less than 100. In binary π = 11·00100 10000 11111 1011 . . . This can be rounded to the repeating decimal 11·00100 1001 . . . , which is equal to $3\frac{1}{7}$.

Ptolemy, the Greek astronomer, used 377/120 (= 3·1416 . . .) but the next great improvement was in China where Tsu Ch'ung-Chi and his son stated that π lay between 3·1415926 and 3·1415927 and gave the approximation 355/113.

Tsu's result was not improved until Al-Kashi in the fifteenth century gave 16 places correctly. European mathematicians at this time were well behind. Fibonacci, for example, found only 3 decimal places correctly.

In the sixteenth century, however, the European mathematicians caught up and then forged ahead. The most successful and the most obsessive was Ludolph van Ceulen who spent much of his life on the calculation of π, first finding it correct to 20 decimal places, then to 32, and finally to 35 places. He did not live to publish his final achievement, but it was engraved on his tombstone in a Leyden church. When the church was rebuilt and his tomb destroyed, his epitaph had already been recorded in a survey of Leyden, and his lifework preserved, but a more lasting monument is the name 'Ludolphian number' which has been used for π in Germany.

About the same time, Adriaen Metius very luckily 'discovered' Tsu's very accurate approximation 355/113, by taking two limits that had actually been calculated by his father, 377/120 and 333/106 and simply averaging the numerators and denominators. This is guaranteed to produce a number lying between the two original fractions, but that is all.

Ludolph's methods were basically the same as Archimedes'. With developments in trigonometry, much superior methods became available. Snell calculated 34 places by using the same geometrical operations that allowed Ludolph to calculate only 14, while Huygens calculated π to 9 places by using only the regular hexagon! Further advances followed rapidly as mathematicians began to understand and use infinite series, limits and the calculus.

None of the calculations of Ludolph or his predecessors had shown any regularity at all in the decimal digits of π. François Viète, the father of modern algebra, showed in 1592 for the very first time a formula for π:

$$\frac{\pi}{2} = \frac{1}{\sqrt{\frac{1}{2}} \sqrt{\frac{1}{2} + \frac{1}{2} \sqrt{\frac{1}{2}}} \sqrt{\frac{1}{2} + \frac{1}{2} \sqrt{\frac{1}{2} + \frac{1}{2} \sqrt{\frac{1}{2}}}} \cdots}$$

A pattern at last! John Wallis followed with:

$$\frac{\pi}{2} = \frac{2}{1} \times \frac{2}{3} \times \frac{4}{3} \times \frac{4}{5} \times \frac{6}{5} \times \frac{6}{7} \times \cdots$$

Isaac Newton, having returned to Grantham in 1666 to escape the Great Plague, easily found π to 16 places using only 22 terms of this series:

$$\pi = \frac{3\sqrt{3}}{4} + 24 \left(\frac{1}{12} - \frac{1}{5 \times 2^5} - \frac{1}{28 \times 2^7} - \frac{1}{72 \times 2^9} - \cdots \right)$$

In 1673 Leibniz discovered that

$$\frac{\pi}{4} = 1 - \frac{1}{3} + \frac{1}{5} - \frac{1}{7} + \frac{1}{9} - \cdots$$

This series is remarkable for its simplicity, but it is hopelessly inefficient as a means of calculating π, because so many hundreds of terms must be calculated to obtain even a few digits of π.

However, by an ingenious sleight of hand, John Machin in 1706 replaced it by a similar formula that allowed him efficiently to calculate to 100 decimal places, far beyond the efforts of Ludolph van Ceulen.

Euler, who first used the Greek letter π in its modern sense, gave an even more impressive demonstration of the power of these new methods by calculating π to 10 decimal places in just one hour.

Euler was a great mathematician, as well as a walking computer. It was he who first revealed the extraordinary relationship between π, *e*, the base of natural logarithms, *i*, the square root of −1, and zero: $e^{i\pi} + 1 = 0$.

Johann Lambert took another significant step forward when he proved that π is irrational. He also calculated, by using continued fractions, the best rational approximations to π, from 103,993/33,102 all the way up to 1,019,514,486,099,146/324,521,540,032,945.

π by this time had long ceased to be merely the ratio of the circumference to the diameter, but the task of simply calculating as many decimal places as possible had not entirely lost its glamour. Indeed, scores of calculations were published. One of the fastest was to 200 places by Johann Dase (1824–1861), completed in less than 2 months.

Dase had been a calculating prodigy as a child and was employed, on the recommendation of Gauss, to calculate tables of logarithms and hyperbolic functions.

In 1853 William Shanks published his calculations of π to 707 decimal places. He used the same formula as Machin and calculated in the process several logarithms to 137 decimal places, and the exact values of 2^{721}.

A Victorian commentator asserted: 'These tremendous stretches of calculation . . . prove more than the capacity of this or that computer for labor and accuracy; they show that there is in the community an increase in skill and courage . . .'

Augustus de Morgan thought he saw something else in Shanks's labours. The digit 7 appeared suspiciously less often than the other digits, only 44 times against an expected average of appearance of 61 for each digit. De Morgan calculated the odds against such a low frequency were 45 to 1. De Morgan, or rather William Shanks, was wrong. In 1945, using a desk calculator, Ferguson found that Shanks had made an error; his calculation was incorrect from place 528 onwards. Shanks, fortunately, was long since dead.

Electronic computers are, of course, vastly superior to human calculators. As early as 1949 the ENIAC calculated π to 2037 places in 70 hours – without making any mistakes. In 1967 a French CDC 6600 calculated 500,000 places, and in 1983 a Japanese team of Yoshiaki Tamura and Tasumasa Kanada produced 16,777,216 (= 2^{24}) places.

What is the point of such calculations? Curiously, it is chiefly to investigate just the kind of irregularities that de Morgan thought he had spotted. It is generally believed that π is normal, and that there is in some sense no pattern at all in the decimal expansion of π, that although it is produced by a definite process, it is effectively random.

It certainly looks random to a rapid examination, despite a chunk of six consecutive 9s between decimal places 762 and 767. Martin Gardner has explained another 'pattern', which occurs much earlier. Here are the 6th to 30th decimal places, slightly spaced to emphasize the pattern:

... 26 53589 793238 46 26 43 383279 ...

A little further on, the 359th, 360th and 361st digits, counting '3' as the first, are 3−6−0, and 315 is similarly centred over the 315th digit.

Such patterns, however, would be expected if π is truly random. Indeed, every possible pattern ought to appear sooner or later. The sequence of digits 123456789 should appear! Does it? No, not so far, apparently, but that is no surprise, because a mere 16,000,000 digits is nothing compared to the endless sequence of digits to come ... However, the pandigital sequence 4592307816 occurs from the 60th to 69th digits.

The first 16 million digits, by the way, have passed all the tests of randomness used on them so far.

What has happened meanwhile to the Greek ambition to square the circle? Several Greek mathematicians thought that they had done so, though their results were at best close approximation.

Mathematicians, not surprisingly, soon learned by experience that the problem was either extraordinarily difficult, or impossible to solve, but their expert opinions had little effect in dampening the ardour of a legion of circle-squarers, some of them exceedingly eminent (in their own, different fields), who could understand the statement of the problem, but not its difficulties.

Nicholas of Cusa (1401–1464) was a cardinal and a famous scholar. He gave 3·1423 as the exact value, but partly redeemed himself by giving a genuinely good trigonometrical approximation, which was later used by Snell. Joseph Scaliger was another notable scholar, a brilliant philologist, with ambitions to be a mathematician, who tried to square the circle. His attempts were refuted by Viète.

Even more curious is the case of the English philosopher Thomas Hobbes (1588–1679) who had learned something of the latest developments in mathematics from Mersenne in Paris. His attempts to square the circle were refuted by John Wallis, whom Hobbes then foolishly attacked. They spent the next quarter-century in bitter argument, doing Wallis no harm at all, but damaging Hobbes's otherwise high reputation.

Jacob Marcelis, in about 1700, supposed that he had squared the circle. His exact value for π was:

$$3\frac{1,008,449,087,377,541,679,894,282,184,894}{6,997,183,637,540,819,440,035,239,271,702}$$

which suggests that he shared some of Shanks's enthusiasm for hard work, without the same justification.

One attempt to square the circle almost reached the statute books. In 1897 House Bill No. 246 was presented to the House of Representatives of the State of Indiana. It was based on the circle-squaring efforts of one Edwin J. Goodwin, a physician but no mathematician, who boldly titled his proposal 'A bill introducing a new mathematical truth'. Despite being both very obscure and very absurd, it sailed through its first reading but was held up before a second reading due to the intervention of C. A. Waldo, a professor of mathematics who happened to be passing through. Its second reading has not taken place to this day!

Such is the pathological self-confidence of many circle-squarers that the breed will no doubt flourish for ever. To mathematicians, however, the problem of squaring the circle was finally answered in 1882 by Lindemann, who proved that π is transcendental, that is, it cannot be the root of any algebraic equation with rational coefficients and only a finite number of terms, more than 80 years after Legendre, having just proved π and π^2 irrational, and reflecting on the history of failure to square the circle, made exactly the same suggestion.

Since every number constructed with ruler and compasses satisfies such an equation, no such construction will ever succeed in squaring the circle. Lindemann's proof, appropriately, used Euler's beautiful relationship.

π has lost some of its mystery, but little of its fascination. It is no longer surprising to find that π appears, for example, in a problem on probability. Count Georges Buffon (1707–1788) the biologist, who also translated Newton on calculus into French, showed that if a needle is dropped from a height randomly on to a parallel ruled surface, the length of the needle equalling the distance between the lines, then the probability that the needle falls across a line is $2/\pi$. Why does π appear in the answer? In this case, because the problem concerns angles, which concern trigonometrical ratios, which concern π . . .

Several investigators have performed experiments to test this conclusion. De Morgan records that one of his pupils made 600 trials and obtained $\pi = 3\cdot137$.

Scores of infinite series involve π in their sums. They are scarcely less beautiful for being well understood. These are as surprising as they are pretty:

$$\frac{\pi\sqrt{2}}{4} = 1 + \frac{1}{3} - \frac{1}{5} - \frac{1}{7} + \frac{1}{9} + \frac{1}{11} - \frac{1}{13} - \frac{1}{15} + \ldots$$

$$\frac{\pi - 3}{4} = \frac{1}{2\times3\times4} - \frac{1}{4\times5\times6} + \frac{1}{6\times7\times8} - \ldots$$

This is a more important result:

$$\frac{\pi^2}{6} = 1 + \frac{1}{2^2} + \frac{1}{3^2} + \frac{1}{4^2} + \ldots$$

If only the odd numbers are used:

$$\frac{\pi^2}{8} = 1 + \frac{1}{3^2} + \frac{1}{5^2} + \frac{1}{7^2} + \frac{1}{9^2} + \ldots$$

Euler first calculated the sums of the even powers of the reciprocals, all the way up to the 26th power:

$$\frac{1}{1^{26}} + \frac{1}{2^{26}} + \frac{1}{3^{26}} + \ldots = \frac{2^{24} \times 76{,}977{,}927 \times \pi^{26}}{27!}$$

$\pi^2/6$ is also equal to this infinite product, through all the primes, also discovered by the prolific Euler:

$$\frac{2^2}{2^2 - 1} \times \frac{3^2}{3^2 - 1} \times \frac{5^2}{5^2 - 1} \times \frac{7^2}{7^2 - 1} \times \frac{11^2}{11^2 - 1} \times \ldots$$

The Indian genius Srinivasa Ramanujan, who had much in common with Euler, produced some extraordinary infinite sums and approximations to π. By a geometrical argument he found

$$\left(9^2 + \frac{19^2}{22}\right)^{\frac{1}{4}} = 3 \cdot 14159\ 26526\ 2 \ldots$$

He also gave

$$\frac{63}{25}(17 + 15\sqrt{5})/(7 + 15\sqrt{5})$$

and the extraordinary

$$2\pi\sqrt{2} = \frac{99^2}{1103}$$

correct to 9 and 8 places respectively.

The most recent methods for calculating π are based on Gauss's study of the arithmetic-geometric mean of two numbers.

Instead of using an infinite sum or product, the calculation goes round and round in a loop. It has the amazing property that the number of correct digits approximately doubles with each circuit of the loop, so that going round a mere 19 times gives π correct to over 1 million decimal places!

Here is a simple loop for calculating π:

The steps must be followed in sequence, up to

$$\frac{(A + B)^2}{4C},$$

[36]

which is the first approximation to π. Then return as the arrow indicates to the first step and go round again. The equals signs stand for 'let —— be ——' rather than equality as in an equation, so the first instruction says 'let Y have the value A'.

$$\left[\begin{array}{l} Y = A \\[2mm] A = \dfrac{A + B}{2} \\[2mm] B = \sqrt{BY} \\[2mm] C = C - X(A - Y)^2 \\[2mm] X = 2X \\[2mm] \text{PRINT } \dfrac{(A + B)^2}{4C} \end{array}\right.$$

The initial values are: $A = X = 1$, $B = 1/\sqrt{2}$, and $C = 1/4$.

Here are the values of π after going round just 3 times on a pocket calculator. It is already correct to 6 decimal places!

loops	approximation to π
1	2·91421 35
2	3·14057 97
3	3·14159 28

3·16227 76601 68379 33199 88935 44432 71853 3719 . . .
$\sqrt{10}$

3·32192 8 . . .
$\log_2 10$

To discover the number of digits of a power of 10, when expressed in binary notation, multiply the index by this number, and take the next highest integer.

Thus $1000 = 10^3$; $3·32192\,8 \ldots \times 3$ is approximately 9·96, so 1000 in base 10 will in binary be of 10 digits. In fact, $1000_{10} = 1,111,101,000_2$.

4
The first composite number, the second square, and the first square of a prime.

The Pythagoreans called numbers divisible by 4, even-even. For this reason, 4, and also 8, were associated with harmony and justice, in contrast to the scales that symbolize justice in modern Western law.

4 is also associated by the Pythagoreans with the tetraktys, the pattern of the first 4 numbers arranged in a triangle. They postulated 4 elements, earth, air, fire and water, symbolized respectively by the cube, octahedron, tetrahedron and icosahedron. The remaining Platonic solid, the dodeca-

hedron, was associated with the sphere of the fixed stars, and later with the quintessence of the medieval alchemists.

A person's temperament was determined by combinations of 4 humours.

Being 2 by 2, there are 4 cardinal points of the compass and 4 corners of the world, and 4 winds.

In the Old Testament there were 4 rivers of paradise, one for each direction, supposed to prefigure the 4 gospels of the New Testament.

The quadrivium of Plato divided mathematics, in his general sense of higher knowledge, into the discrete and the continuous. The absolute discrete was arithmetic, the relative discrete was music. The stable continuous was geometry and the moving continuous, astronomy.

The most pleasing musical intervals are associated with the ratios of the numbers 1 to 4.

The Greeks also associated 4 with solid objects, notwithstanding their association between 3 and volume. They followed the natural progression, 1 for a point, 2 for a line, 3 for a surface, and 4 for a solid.

The simplest Platonic solid, the tetrahedron, has 4 vertices and 4 faces.

A square has 4 edges and 4 vertices. A cube has square faces, while its dual, the octahedron, has 4 faces about each vertex.

Being 2^2, a plane figure with bilateral symmetry about two different lines is divided into a 4 congruent parts.

Einstein's space-time is 4-dimensional. However, in recent theories, 4 dimensions are insufficient.

A hyperbola can be drawn through any 4 points in the plane, no 3 of which are collinear.

Every integer is the sum of at most 4 squares. This celebrated theorem may have been known empirically to Diophantus. Bachet tested it successfully up to 120 and stated it in his edition of Diophantus, to which he added some of his own material. It was studied by Fermat and Euler, who failed to solve it, and finally proved by Lagrange in 1770. Only one-sixth of all numbers, those of the form $4^n(8m + 7)$, however, actually require 4 squares. The remainder are the sum of at most 3 squares.

Ferrari first solved equations of the 4th degree. His solution was published by Cardan in his *Ars Magna*. The general equation of higher degree cannot be solved by the use of radicals.

The 4-colour problem

For more than a century the 4-colour conjecture was one of the great unsolved problems of mathematics. Some mathematicians would still say that it has not been solved satisfactorily.

In October 1852, Francis Guthrie was colouring a map of England. It suddenly occurred to him to wonder how many colours were needed if, as is natural, no two adjacent counties were given the same colour. He supposed the answer was 4. It was published in 1878, setting in motion a bizarre but not untypical sequence of events.

Kempe thought that he had proved it in 1879, but 11 years later his proof was shown to be faulty. Meanwhile, in 1880, the conjecture had been proved again, but this proof was also flawed. However, these attempts were valuable in deepening mathematicians' understanding of the problem. Indeed, many important concepts in graph theory were developed through attacks on this problem, which however proved extremely resistant. The solution was finally achieved in 1976 by Wolfgang Haken and Kenneth Appel who transformed the problem into a set of sub-problems that could be checked by computer.

Mathematicians have been sceptical because of the lengthy reasoning involved, and the length of time, 1200 hours, taken on the computer. The very existence of a proof that few other mathematicians will ever be able to check is a recent development in mathematics. Another example of the same phenomenon is the classification of finite groups. This classification is now complete but the entire proof is spread across thousands of pages in different journals published over many years. This contradicts the traditional idea of a proof as an available means of confirming a thesis and persuading others also that it is true.

4 is exceptional in not dividing $(4 - 1)! = 3!$. It is the only composite n which does not divide $(n - 1)!$.

Brocard's problem asks: When is $n! + 1$ a square? $4! + 1 = 5^2$.

A number is divisible by 4 if the number represented by its last 2 digits is divisible by 4.

Starting with any number, form a new number by adding the squares of its digits. Repeat. This process eventually either sticks on 1, or goes round a loop of which 4 is the smallest member: $4 - 16 - 37 - 58 - 89 - 145 - 42 - 20 - 4 \ldots$

If a number in base 10 is a multiple of its reversal (numbers beginning with 0 are excluded), their ratio is either 4 or 9.

4 is the only number equal to the number of letters in its normal English expression: 'four'.

4·12310 5 . . .
$\sqrt{17}$, the highest root to be proved irrational by Theodorus.

4·66920 16609 0 . . .
The Feigenbaum constant.

[39]

5

The Pythagoreans associated the number 5 with marriage, because it is the sum of what were to them the first even, female number, 2, and the first odd, male number, 3.

5 is the hypotenuse of the smallest Pythagorean triangle, that is, a right-angled triangle with integral sides.

The Pythagoreans also associated this triangle with marriage and Pythagoras' theorem was sometimes called the Theorem of the Bride. The sides 3 and 4 were associated with the male and female respectively, and the hypotenuse, 5, with the offspring.

The 3–4–5 triangle is the only Pythagorean triangle whose sides are in arithmetical progression, and the only one whose area is one-half of its perimeter.

The mystic pentagram, which was so important to the Pythagoreans, was known in Babylonia and probably imported from there. The Penta-gram was associated with the division of line in extreme and mean proportion, the Golden Section, and also with the 4th of the regular solids, the dodecahedron, whose faces are regular pentagons. The early Pythagoreans did not know the 5th regular Platonic solid, the icosahedron. By constructing a nest of pentagrams inside a regular pentagon, it is relatively easy to show that subtraction of the sides and diagonals can be continued indefinitely. It has been suggested that this pattern led to the idea that some lengths are incommensurable.

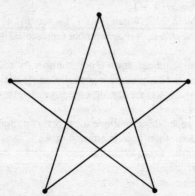

The Pythagoreans, according to Plutarch, also called 5 nature, because when multiplied by itself, it terminates in itself. That is, all powers of 5 end in the digit 5. They knew that 6 shares this property, but no other digit. In modern terminology, 5 and 6 are the smallest automorphic numbers.

5 is the sum of 2 squares, $5 = 1^2 + 2^2$, like any hypotenuse of a Pythagorean triangle.

It is also a prime, the first, of the form $4n + 1$, from which it follows that it is the sum of 2 squares in one way only.

5 is the 1st prime of the form $6n - 1$. All primes are one more or one less than a multiple of 6, except 2 and 3.

Pappus showed how to construct a conic through any 5 points in the plane, no 3 of which are collinear.

5 is the 2nd Fermat number and the 2nd Fermat prime; $5 = 2^2 + 1$. Only 5 Fermat primes are known to exist.

The 5th Mersenne number, $2^5 - 1 = 31$ and is prime, the 3rd to be so, leading to the 3rd perfect number, 496.

$5! + 1$ is a square.

Every number is the sum of 5 positive or negative cubes in an infinite number of ways.

The general algebraic equation of the 5th degree cannot be solved in radicals. First proved by Abel in 1824.

Lamé showed that the Euclidean algorithm for finding the highest common factor of 2 numbers takes in base 10 at most 5 times as many steps as there are digits in the smallest number.

5 is a member of 2 pairs of twin primes, 3 and 5, and 5 and 7.

$5-11-17-23$ is the smallest sequence of 4 primes in arithmetical progression. Add the prime 29 to form the smallest set of 5 primes in arithmetical progression.

5 is probably the only odd untouchable number.

The volume of the unit 'sphere' in hyperspace increases up to 5-dimensional space, and decreases thereafter.

Counting in 5s
This might seem a natural base for a counting system, since we have 5 fingers per hand. However, only one language uses a counting system based exclusively on 5, Saraveca, a South American Arawakan language, though 5 has a special significance in many counting systems based on 10 and 20. For example in many Central American languages, the numbers 6 through 9 are expressed as $5 + 1$, $5 + 2$ and so on.

The Romans used $V = 5$, $L = 50$ and $D = 500$, so 664 was DCLXIIII. (The idea of placing an I before V to represent 4, or I before X for 9, for example, which makes numbers shorter to write while making them more confusing for arithmetic, was hardly ever used by the Romans themselves and became popular in Europe only after the invention of printing.)

Divisibility

Because 5, like 2, is a factor of 10, decimal fractions such as 1/20, whose denominators are products of 2s and 5s only, have finite decimal expansions and do not recur. More precisely, if $n = 2^p 5^q$, then the length of $1/n$ as a decimal is the greater of p and q.

If $1/m$ is a recurring decimal, and $1/n$ terminates, then $1/mn$ has a nonperiodic part whose length is that of $1/n$, and a recurring part whose length is the period of $1/m$.

The Platonic solids

There are 5 Platonic solids, the regular tetrahedron, cube, octahedron, dodecahedron and icosahedron, all but the cube being named after the Greek word for their number of faces.

They were all known to the Greeks, Theaetetus, a pupil of Plato, showed how to inscribe the last 2 in a sphere. Euclid showed, by considering the possible arrangements of regular polygons around a point, that there are no more than 5.

Kepler used them, with typical confidence in their mystical properties, to explain the relative sizes of the orbits of the planets:

The earth's orbit is the measure of all things; circumscribe around it a dodeca-hedron, and the circle containing this will be Mars: circumscribe around Mars a tetrahedron, and the circle containing this will be Jupiter: circumscribe around Jupiter a cube, and the circle containing this will be Saturn. Now inscribe within the earth an icosahedron, and the circle contained in it will be Venus; inscribe within Venus an octahedron, and the circle contained in it will be Mercury. You now have the reason for the number of planets.

The idea of a polyhedron can be extended to more than 3 dimensions, just as a polyhedron can be considered as a 3-dimensional polygon.

There are 5 cells in the simplest regular 4-dimensional polytope, called the simplex, which also has 10 faces, 10 edges and 5 vertices, so that it is self-dual.

If the Riemann Hypothesis is true, then every odd integer, apart from 1, can be written as the sum of at most 5 primes. [*Acta Arithmetica* v72 361]

$n!$ never ends in 5 zeros. The sequence of such numbers of zeros continues: 11 17 23 29 30 36 42 . . . [Sloane 3808 and MM v27 53]

The Fibonacci sequence

5 is the 5th Fibonacci number. Leonardo of Pisa, called Fibonacci, discussed in his *Liber Abaci* this problem: A certain man put a pair of rabbits in a place surrounded on all sides by a wall. How many pairs of rabbits can be produced from that pair in a year if it is supposed that

every month each pair begets a new pair which from the second month on becomes productive?

Assuming that the rabbits are immortal, the number at the end of each month follows this sequence. (Leonardo omitted the first term, supposing that the first pair bred immediately.)

1 1 2 3 5 8 13 21 34 55 89 144 233 . . .

It was christened the Fibonacci sequence by Edouard Lucas in 1877, when he used it, and another sequence now named after himself, to search for primes among the Mersenne numbers. It is one of the curious coincidences that occur in the history of mathematics that a problem about rabbits should generate a sequence of numbers of such interest and fascination. Rabbits, needless to say, do not feature again in its history.

Its first and simplest property is that each term is the sum of the two previous terms. Thus the next term will be $144 + 233 = 377$. This was surely known to Fibonacci, though he nowhere states it. Mathematicians do not always state the obvious.

Kepler believed that almost all trees and bushes have flowers with 5 petals and consequently fruits with 5 compartments. He naturally associated this fact with the regular pentagon and the Divine Proportion. He continues:

It is so arranged that the two lesser terms of a progressive series added together constitute the third . . . and so on to infinity, as the same proportion continues unbroken. It is impossible to provide a perfect example in round numbers. However . . . Let the smallest numbers be 1 and 1, which you must imagine as unequal. Add them, and the sum will be 2: add to this 1, result 3; add 2 to this, and get 5; add 3, get 8 . . . As 5 is to 8, so 8 is to 13, approximately, and as 8 is to 13, so 13 is to 21, approximately.

This statement could scarcely be clearer, but it was not until 1753 that the Scottish mathematician Robert Simson first stated explicitly that the ratios of consecutive terms tend to a limit, which is ϕ, the Golden Ratio. These are the first few ratios: 1/1 2/1 3/2 5/3 8/5 13/8 21/13 34/21 55/34 89/55 144/89 233/144 . . .

Successive ratios are alternately less than and greater than the Golden Ratio. After 12 terms the match with ϕ is correct to 4 decimal places. For much higher values the Fibonacci sequence matches the geometric sequence ϕ^n very closely indeed. (This is a consequence only of the rule that each term is the sum of the 2 preceding terms. Start with any 2 integers, construct a generalized Fibonacci series, by adding successive terms to get the next, and their ratio will tend to ϕ.)

More precisely, as Euler knew, and Binet rediscovered in 1843 the nth Fibonacci number is given by the formula:

$$F_n = \frac{(1 + \sqrt{5})^n - (1 - \sqrt{5})^n}{2^n \times \sqrt{5}}$$

There is another version of this formula, which is simpler to use in practice. Because

$$\frac{1}{\sqrt{5}}\left(\frac{1 + \sqrt{5}}{2}\right)^n$$

is only $0.618\ldots$ when $n = 1$ and rapidly becomes very small indeed, F_n is actually the nearest integer to

$$\left(\frac{\sqrt{5} - 1}{2}\right)^n$$

For example F_8 is the integral part of $21.00952\ldots$ which is 21.

Simson also discovered the identity $F_{n-1}F_{n+1} - F_n^2 = (-1)^n$, which is the basis for a puzzling trick first presented by Sam Loyd.

Draw on graph paper a square whose side is a Fibonacci number with even subscript, say 8. Divide it as shown, and the pieces can be reassembled to form a rectangle with area 65. Where has the extra square come from?

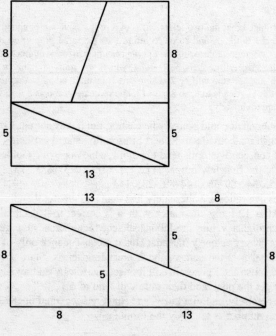

Nowhere, of course. The diagonal of the second figure is actually 2 halves of a long thin parallelogram, with area 1 unit. It seems to be a genuine straight line only because the slopes of the two sides, 3/8 and 2/5, are so similar. If we had started with a higher Fibonacci number, say 21, the illusion would be even closer and even more convincing.

The number of Fibonacci identities is literally endless. Lucas discovered a relationship between Fibonacci numbers and the binomial coefficients:

$$F_{n+1} = \binom{n}{0} + \binom{n-1}{1} + \binom{n-2}{1} + \ldots$$

For example:

$$F_{12} = 144 = \binom{11}{0} + \binom{10}{1} + \binom{9}{2} + \binom{8}{3} + \binom{7}{4} + \binom{6}{5}$$
$$= 1 + 10 + 36 + 56 + 35 + 6$$

Catalan showed a similar result:

$$2^{n-1} F_n = \binom{n}{1} + 5\binom{n}{3} + 5^2\binom{n}{5} + \ldots$$

The sums of the first n terms, and of the terms with even subscripts and odd subscripts, can all be expressed very neatly:

$$F_1 + F_2 + F_3 + F_4 + \ldots F_n = F_{n+2} - 1$$
$$F_1 + F_3 + F_5 + F_7 + \ldots F_{2n-1} = F_{2n}$$
$$F_2 + F_4 + F_6 + F_8 + \ldots F_{2n} = F_{2n+1} - 1$$

Similarly, $F_1^2 + F_2^2 + F_3^2 + \ldots + F_n^2 = F_n F_{n+1}$, which can be illustrated nicely in a figure, which naturally is almost identical to the figure on page 22: the proportions of this figure, $55 : 34$, are already a fair approximation to ϕ.

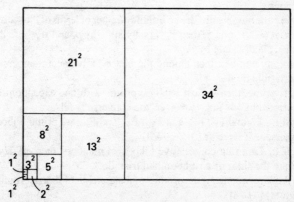

There are many more identities similar to Simson's, such as:

$$F_{2n} = F_{n+1}^2 - F_{n-1}^2 \quad \text{or} \quad F_{3n} = F_{n+1}^3 + F_n^3 - F_{n-1}^3$$

Charles Raine ingeniously connected Fibonacci numbers to Pythagorean triangles. Take any 4 consecutive Fibonacci numbers; the product of the outer terms and twice the product of the inner terms are the legs of a Pythagorean triangle: for example, 3, 5, 8, 13, gives the 2 legs, 39 and 80, of the right-angled triangle 39–80–89. The hypotenuse, 89, is also a Fibonacci number! Its subscript is half the sum of the subscripts of the 4 original numbers. Finally, the area of the triangle is the product of the original 4 numbers, 1560.

(Incidentally, no four terms of the Fibonacci sequence can be in arithmetic progression.)

The sums of the 2 series

$$\frac{1}{1 \times 2} - \frac{1}{2 \times 3} + \frac{1}{3 \times 5} - \frac{1}{5 \times 8} \ldots$$

and

$$\frac{1}{1 \times 3} + \frac{1}{3 \times 8} + \frac{1}{8 \times 21} + \frac{1}{21 \times 55} + \ldots$$

are equal to ϕ^{-2}. [Pincus Schub]

The number of Fibonacci numbers between n and $2n$ is either 1 or 2, and the number of Fibonacci numbers having the same number of digits is either 4 or 5. [K. Subba Rao]

The Fibonacci numbers possess very elegant divisibility properties. Consider two numbers, m and n. If m divides n, then F_m divides F_n. If the highest common factor of p and q is r, then the highest common factor of F_p and F_q is F_r. It follows that any two consecutive Fibonacci numbers are coprime.

Every prime number divides an infinite number of terms of the sequence. In fact, if $p = \pm 1$ mod 5, then F_{p-1} is divisible by p, and if $p = \pm 2$ mod 5, then F_{p+1} is divisible by p.

If m is any number, then among the first m^2 Fibonacci numbers there is one divisible by m.

If F_n is prime, then n must itself be prime, with one exception: $F_4 = 3$, 3 is prime but 4 is not. However, the converse is false.

F_5 and F_7, F_{11} and F_{13}, F_{431} and F_{433}, F_{569} and F_{571} are prime, Fibonacci prime pairs, as it were. [MOC v50 251]

If a, b, c, d are four consecutive Fibonacci numbers, then ad, $2bc$ and $b^2 + c^2$ are the sides of a right-angled triangle.

No Fibonacci number is the product of just 2 smaller Fibonacci numbers. [Stephany, MM v66 61]

The sequence of numbers, n, such that F_n ends in n, starts: 1 5 25 29 41 49 . . . [FQ v4 156]

The Fibonacci sequence is also linked in a surprising way with the growth of plants. Kepler may have realized this. He writes:

It is in the likeness of this self-developing series that the faculty of propagation is, in my opinion, formed; and so in a flower the authentic flag of this faculty is shown, the pentagon. I pass over all the other arguments that a delightful rumination could adduce in proof of this.

What were Kepler's other arguments? He does not say, but in the nineteenth century Schimper and Braun investigated phyllotaxis, the arrangements of leaves round a stem. Leaves grow in a spiral, such that the angles between each pair of successive leaves are constant. The commonest angles are 180°, 120°, 144°, 135°, 138°27', 137°8', 137°38', 137°27', 137°31' . . . which seem to be tending to a limit.

What that limit is becomes clearer when they are expressed as ratios of a complete circle. These ratios are, respectively, 1/2, 1/3, 2/5, 3/8, 5/13, 8/21, 13/34, 21/55 and 34/89, the ratios of alternate members of the Fibonacci sequence.

To put that another way, the numerator and denominator of each new fraction are sums of the numerators and denominators of the previous two fractions. These ratios tend to the limiting value ϕ^{-2}, and the limiting angle is approximately 137°30' and 28 seconds, which divides the circumference of a circle in the Golden Ratio.

The smallest ratios, 1/2 and 1/3, are found in grasses and sedges but are otherwise not very common, though commoner than 1/4 and 1/5, which do not exist and form part of another Fibonacci-type sequence. The most frequent leaf arrangements are 2/5, found in roses, and 3/8. Much higher ratios, however, appear much more clearly in the scales of a fir cone or the florets of a sunflower, which are packed closely together. The packing is highly regular, forming sets of spiral rows, or parastichies, two of which are more prominent than the rest.

A pineapple usually has 8 and 13 parastichies. A sunflower may have from 21/34 up to as high as 89/144. Even 144/233 has been claimed for one giant plant. (Although plants of the same species and even of the same family tend to have the same parastichy numbers, the higher numbers especially do vary from plant to plant. The phyllotaxis may even change as a plant grows, starting with a low ratio such as 1/2 or 1/3 and then changing to higher ratios.)

Why do plants grow this way? Less entranced by the Fibonacci numbers than mathematicians, botanists are more interested in an explanation, on which they do not yet agree. One plausible theory, which might be explained by chemical inhibition of growth, is that each primordium, the primitive leaf bud, develops in the largest gap available. Whatever the botanists eventually decide, mathematicians will continue to delight in this connection between rabbits and the plants they eat.

The Fibonacci numbers have other uses in more advanced mathematics. The Russian Matijasevic used Fibonacci to finally solve Hilbert's 10th problem. No algorithm exists that, given any Diophantine equation, will decide within a finite number of steps whether it has a solution. He exploited the rate at which the sequence of Fibonacci numbers increases. They have also recently found further uses in computer science, in designing efficient algorithms for constructing and searching tables of data, for example. [Kepler, *The Six-cornered Snowflake*, OUP, 1966]

5·25694 64048 60 . . .

The approximate 'volumes' of the unit radius 'spheres' in dimensions from 1 upwards are:

dim.1	dim.2	dim.3	dim.4	dim.5	dim.6	dim.7	...
2	3·1	4·2	4·9	5·264	5·2	4·7	...

The volume is a maximum in 5 dimensions, and declines thereafter. If however the dimension is regarded as a real variable, able to take non-integral values, then the maximum volume occurs in 'space' of this dimension, 5·256 . . . The volume is then 5·27776 8 . . . compared to the volume in 5 dimensions of 5·26378 9 . . . [David Singmaster]

6

The second composite number and the first with 2 distinct factors.

Therefore the first number, apart from 1, which is not the power of a prime.

The Pythagoreans associated 6 with marriage and health, because it is the product of their first even and first odd numbers, which were female and male respectively. It also stood for equilibrium, symbolized by 2 triangles, base to base.

It is the area and the semi-perimeter of the first Pythagorean triangle, with sides, 3, 4, 5.

The first perfect number, as defined by Euclid. Its factors are 1, 2, 3 and 6 = 1 + 2 + 3.

It is the only perfect number that is not the sum of successive cubes.

St Augustine wrote, 'Six is a number perfect in itself . . . God created all things in six days because this number is perfect. And it would remain perfect, even if the work of six days did not exist.' [Bieler]

6 is also equal to $1 \times 2 \times 3$, and is therefore the 3rd factorial, 3!, and also the second primorial.

No other number is the product of 3 numbers and the sum of the same 3 numbers.

1, 2, 3 is also the only set of 3 integers such that each divides the sum of the other 2.

6 also equals $\sqrt{1^3 + 2^3 + 3^3}$.

The equation $x^3 + y^3 + z^3 = 6xyz$ has the unique solution $x = 1$, $y = 2$, $z = 3$. [*Acta Arithmetica* v73 201]

It is the only number that is the sum of exactly 3 of its factors, which is the same as saying that 1 can be expressed uniquely as the sum of 3 unit fractions, the smallest of which is $1/6$: $1 = 1/2 + 1/3 + 1/6$.

6^2 ends in 6. The other digit with this property is 5.

Every prime number greater than 3 is of the form $6n \pm 1$.

Any number of the form $6n - 1$ has two factors whose sum is divisible by 6.

6 is the 3rd triangular number, and the only triangular number, apart from 1, with less than 660 digits whose square (36) is also triangular.

The only consecutive integers whose sums of the squares of their divisors are equal are 6 and 7. [Guy 68]

The number of Goldbach decompositions of an even number, N, into the sum of two primes have local maxima when N is a multiple of 6. [Sternheimer, JRM v24 30]

The following property is due to Iamblichus. Take any 3 consecutive numbers, the largest divisible by 3. Add them, and add the digits of the result, repeating until a single number is reached. That number will be 6.

The 2nd and 3rd Platonic solids, which are duals of each other, the cube and the octahedron, have 6 faces and 6 vertices respectively.

The first, the tetrahedron, has 6 edges.

Regular polytopes

There are 6 regular polytopes. They are the analogues in 4 dimensions of the regular polyhedra in 3 dimensions and the regular polygons in 2 dimensions.

Each polytope has vertices, edges, faces and also cells. Two of them are self-dual; the others form 2 dual pairs.

name	number of cells	number of faces	number of edges	number of vertices
simplex	5	10	10	5
tesseract	8	24	32	16
16-cell	16	32	24	8
24-cell	24	96	96	24
120-cell	120	720	1200	600
600-cell	600	1200	720	120

6 equal circles can touch another circle in the plane.

One of the 3 regular tessellations of the plane is composed of regular hexagons.

Pappus discussed the practical intelligence of bees in constructing hexagonal cells. He supposed that the cells must be contiguous, to allow no foreign matter to enter, must be regular, and therefore either triangular, square, or hexagonal, and concluded that bees knew that a hexagon, using the same material, would hold more than the other shapes. Pappus, claiming that man has a greater share of wisdom than the bees, then went on to show that of all regular figures with equal perimeter, the one with the larger number of sides has the larger area, the circle being the limiting maximum.

Kepler discussed the 6-fold symmetry of snowflakes, and attempted to explain it by considering the close packing of spheres in a hexagonal array.

Pascal discovered in 1640 at the age of 16 his theorem of the Mystic Hexagram. If any six points are chosen on a conic section, labelled 1, 2, 3, 4, 5, 6, then the intersections of the lines 12 and 45, 34 and 61, 56 and 23, will lie on a straight line.

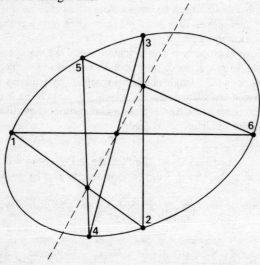

Brianchon enunciated the dual theorem, in which the 6 original points are replaced by 6 tangents to the conic.

6·196 . . .

The approximate length of the shortest Steiner tree connecting the vertices of a unit cube. [MG v78 161]

6·28318 5 . . .

2π

The ratio of the circumference to a radius of a circle. The number of radians in a complete circle.

7

7 days in a week, and therefore associated with 14 and with 28 days in a lunar month.

The 4th prime number, and the first of the form $6n + 1$.

The start of an arithmetical progression of six primes: 7, 37, 67, 97, 127, 157.

7 and 11 are the first pair of consecutive primes different by 4.

The 3rd Mersenne number, $7 = 2^3 - 1$, and the second Mersenne prime, leading to the 2nd perfect number.

The 1st number that is not the sum of at most 3 squares. The sequence of such numbers continues: 15   23   28   31   39   47   55   60   . . .

$7 = 3! + 1$. $n! + 1$ is prime for $n = 1, 2, 3, 11, 27, 37, 41, 73, 77, 116, 154, 320, 340, 399, 427$, and no other values below 546.

Brocard's problem. When is $n! + 1$ a square? The only known solutions are $n = 4, 5$ and 7: $7! + 1 = 5041 = 71^2$.

The Fermat quotient

$$\frac{2^{p-1} - 1}{p}$$

is a square only when p is 3 or 7.

Lamé proved in 1840 that Fermat's equation $x^7 + y^7 = z^7$ has no solutions in integers.

If a, b are the shorter sides of a Pythagorean triangle, then 7 divides one of $a, b, a - b$ or $a - b$.

Because 7^2 falls short of 50 by only 1, 7 was called by the Greeks, the *rational diagonal* of a square of side 5.

All sufficiently large numbers are the sum of 7 positive cubes.

To test if a number is divisible by 7: multiply the left-hand digit by 3 and add the next digit. Multiply the answer by 3. Repeat as often as necessary. If the final answer is divisible by 7, so is the original number. Alternatively, start by multiplying the right-hand digit by 5 and adding the adjacent digit. Repeat as before.

7 numbers are sufficient to colour any map on a torus. Surprisingly, this was known before the 4-colour conjecture was solved for plane maps.

At least 7 rectangles are required if a rectangle is to be divided into smaller rectangles no one of which will fit inside another. The smallest rectangle that can be tiled 'incomparably' is 13 by 22. [Yao and Reingold, JRM v8]

At least 7 rectangles are also required to divide a rectangle into smaller rectangles of different shape but equal area.

An obtuse-angled triangle can be divided into not less than 7 acute-angled triangles.

There are 7 basically different patterns of symmetry for a frieze design.

The regular 7-gon is the smallest that cannot be constructed by ruler and compass alone.

7 is the smallest prime the period of whose reciprocal in base 10 has maximum length. $1/7 = 0 \cdot 14285\ 71428\ 57 \ldots$ (*See 142,857*.)

The unique projective plane of order 2, called the Fano plane, has 7 lines and 7 points, with 3 points on each line and 3 lines through each point.

There are 7 prime knots with 7 crossings. (The product of two knots is found by tying them on the same string. A prime knot is not a product.)

It takes 7 riffle shuffles to randomize a pack of cards. [Diaconis and Bayer, *Discover*, Jan. 1991]

Three odd squares, 5^2, 11^2 and 181^2 are 7 less than the nearest power of 2. [*Acta Arithmetica* v38 409]

7 and 8 are successive solutions of $\phi(n) = \phi(n + 2)$. $\phi(7) = \phi(9) = 6$ and $\phi(8) = \phi(10) = 4$.

$7^2 + 8^2 + \ldots + 28^2 + 29^2$, and $7^2 + 8^2 + \ldots + 38^2 + 39^2$, and $7^2 + 8^2 + \ldots + 55^2 + 56^2$, and $7^2 + 8^2 + \ldots + 189^2 + 190^2$ are all examples of a sequence of consecutive squares, whose sums are also squares. [Beeckmans, AMM v101 439]

The problem of St Ives

This Mother Goose rhyme is well known:

'As I was going to St Ives, I met a man with seven wives. Every wife had seven sacks, and every sack had seven cats, every cat had seven kittens. Kittens, cats, sacks and wives, how many were going to St Ives?'

Problem 79 of the Rhind papyrus, written by the scribe Ahmes, which dates from about 1650 BC, concerns:

Houses	7
Cats	49
Mice	343
Spelt	2401
Hekat	16807
TOTAL	19607

The resemblance is remarkable. Moreover, there is a connecting link, of sorts. Leonardo of Pisa, called Fibonacci, in his *Liber Abaci* (1202 and 1228) also includes the same problem. Pierce comments that it seems to be of the same origin as the House that Jack built, and that Leonardo uses the same numbers as Ahmes and makes his calculations in the same way. It is tempting to suppose that this problem is indeed more than 3500 years old, and has survived essentially unchanged throughout that time.

[Gillings, *Mathematics in the Time of the Pharaohs*, MIT Press, 1972; and Eisele in 'Liber Abaci', *Scripta Mathematica*, v17]

8

The second cube: $8 = 2^3$. The only cube that is 1 less than a square $8 = 3^2 - 1$ and the only power that differs by 1 from another prime power.

The 6th Fibonacci number, and the only Fibonacci number that is a cube, apart from 1.

The number of parts into which 3-dimensional space is divided by 3 general planes.

There are 8 notes in an octave.

It is possible to place the maximum 8 queens on a chessboard, so that no queen attacks any other, in 12 essentially different ways.

8 times any triangular number is 1 less than a square.

A number is divisible by 8 if the number formed by its last 3 digits is divisible by 8.

The second octagonal number, given by the formula $n(3n - 2)$. The sequence starts: 1 8 21 40 65 96 . . .

Magic cubes

Perfect magic cubes, in which all the rows, columns and diagonals of every layer, plus the space diagonals through the centre, sum to the same total are impossible for orders 3 ($3 \times 3 \times 3$) and 4 ($4 \times 4 \times 4$). It is not known if such cubes can exist for orders 5 and 6.

Magic cubes do exist for order 8. The first was privately published in 1905, a method of construction was again discovered in the late 1930s, and in 1976 Martin Gardner published an example constructed by Richard Myers. Myers discovered how to construct vast numbers of them by superimposing 3 Latin cubes and using octal notation, when he was a 16-year-old schoolboy.

Soon after Gardner reported on Myers's discovery, Richard Schroeppel and Ernst Straus independently found order-7 magic cubes.

[Gardner, *Scientific American*, Jan. 1976]

The octal system

8 is the base of the octonary, octenary, or octal system. Emmanuel Swedenborg, the Danish philosopher, wrote a book advocating base 8. It has much of the simplicity of the binary system. All its factors are powers of 2, yet numbers of a reasonable size do not take an absurdly large number of digits to express. 100 in base 10 is 144 in base 8 and 1,100,100 in binary. The binary is much harder to remember (always a great disadvantage for practical purposes) and longer, though it can be obtained instantly from the octal 144 by replacing the digits by their binary expression. 1–4–4 becomes 1–100–100 or 1,100,100.

Arguments for changing to base 8 completely are weaker than for changing to duodecimal. But because of the connection with binary, it has been used extensively in computers, though since the IBM 360 series was introduced in the early 1960s, using base 16 (hexadecimal), it has fallen out of favour.

A deltahedron is a polyhedron all of whose faces are triangular. There are an unlimited number of them, since any deltahedron has exposed faces to which another triangular pyramid can be attached. However, only 8 of them are convex. 3 of these are the regular tetrahedron, octahedron and icosahedron. 2 more are a pair of tetrahedra glued face to face, and a pair of pentagonal pyramids glued face to face. The octahedron has 8 triangular faces, and 6 vertices and 12 edges, making it the dual of the cube, which has 8 vertices, 6 faces and 12 edges. Thus, if the 6 mid-points of the faces of a cube are joined together, they form an octahedron. Conversely, the 8 mid-points of the faces of an octahedron join to form a cube.

9

The 3rd square, and therefore the sum of 2 consecutive triangular numbers: $9 = 3 + 6$.

Written as '100' in base 3.

The first odd prime power, and with 8 the only powers known to differ by 1.

The only square that is the sum of 2 consecutive cubes: $9 = 1^3 + 2^3$.

The 4th Lucky number, and the 1st square Lucky number apart from 1.

$9 = 1! + 2! + 3!$

The smallest Kaprekar number apart from 1: $9^2 = 81$ and $8 + 1 = 9$.

9 is subfactorial 4.

There are 9 regular polyhedra, the 5 Platonic solids and the 4 Kepler–Poinsot stellated polyhedra.

9 is the smallest number of distinct integral squares into which a rectangle may be divided. The smallest solution is 32 by 33 and the squares have sides 1, 4, 7, 8, 9, 10, 14, 15, and 18.

The Feuerbach, or nine-point circle of a triangle

In 1820 Brianchon and Poncelet proved that the feet of the altitudes, the mid-points of the sides and the mid-points of the segments of the altitudes from the vertices to their point of intersection, all lie on a circle.

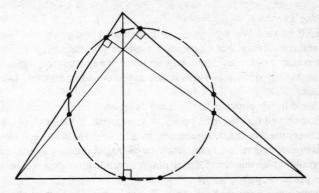

Feuerbach proved 2 years later that this circle also touches the inscribed and 3 escribed circles of the triangle, and in consequence it is often known as the Feuerbach circle.

Because 9 is 1 less than the base of our usual counting system, there is a simple test for divisibility by 9. 9 divides a number if and only if it divides the sum of the number's digits.

Arithmetical sums may be checked by the process called 'casting out nines'. This came to Europe from the Arabs, but was probably an Indian invention. Leonardo of Pisa described it in his *Liber Abaci*. Each number in a sum is replaced by the sum of its digits. (Originally it was replaced by the remainder on dividing by 9, which is a long way round of coming to the same result.) If the original sum is correct, so will the same sum be when performed with the sums-of-digits only,.

Which fits better, a round peg in a square hole or a square peg in a round hole? This can be interpreted as, which is larger, the ratio of the area of a circle to its circumscribed square, or the area of a square to its circumscribed circle? In 2 dimensions, these ratios are $\pi/4$ and $2/\pi$ respectively, so a round peg fits better into a square hole than a square peg fits into a round hole.

However, this result is true only in dimensions less than 9. For $n \geq 9$ the n-dimensional unit cube fits more closely into the n-dimensional unit sphere than the other way round. [Singmaster, MM v37]

There are no configurations of 7 or 8 lines such that there are 3 points on each line and three lines through each point that can actually be realized geometrically. There are 3 essentially different such configurations with 9 lines. The first of these is the configuration of Pappus's theorem, which is a special case of Pascal's Mystic Hexagram.

Waring's problem

In 1770 Edward Waring wrote in his *Meditationes algebraicae*, 'Every integral number is either a cube, or is a sum of two, three, 4 ,5, 6, 7, 8 or nine cubes; it is furthermore a biquadrate or is a sum of two, three, etc., all the way up to nineteen biquadrates, and so on in like manner.' [*Scripta Mathematica* v7]

This difficult problem has still not been completely solved, though Hilbert proved that for each power, k, there exists a number, $g(k)$, such that every number can be represented by at most $g(k)$ kth powers. Not all numbers, of course, are 'sufficiently large' and it remains a problem to determine which numbers for each power k require more than $g(k)$ powers to represent them.

Waring was correct about cubes, though only a finite set of numbers actually requires 9, and he was right about 4th powers, though again, 19 is more than sufficient for all but a finite set of numbers.

Magic squares

The first 9 numbers can be arranged in a magic square so that all rows, columns and both diagonals have the same sum, 15. This can be done in essentially only one way, all solutions being related by reflections and rotations to each other. The illustration on p. 57 is the Lo Shu, the magic square as it was known to the ancient Chinese.

This pattern has other beautiful properties. The number 5, halfway between 1 and 9, naturally occupies the middle cell.

All four lines through the central 5 are in arithmetical progression, with differences 1, 2, 3, 4 rotating anti-clockwise from $6-5-4$ to $9-5-1$.

The sums of the squares of the 1st and 3rd columns are equal: $4^2 + 3^2 + 8^2 = 2^2 + 7^2 + 6^2 = 89$. The middle column gives $9^2 + 5^2 + 1^2 = 107 = 89 + 18$.

The squares of the numbers in the rows sum to 101, 83 and 101, and $101 - 83 = 18$.

$492^2 + 357^2 + 816^2 = 294^2 + 753^2 + 618^2$, and the same pattern is true for the columns, and for the diagonals (456, 978, 231).

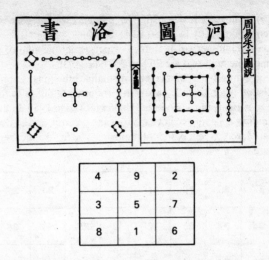

4	9	2
3	5	7
8	1	6

There are just 8 ways in which the magic total 15 can be made by adding 3 of the integers 1 to 9. Each of these 8 ways occurs once in the square.

9·86960 44010 89358 . . .

π^2. Legendre proved in 1794 that π^2 is irrational.

10

The base of our counting system, it therefore has the simplest test of divisibility. The number of consecutive zeros, counting from the units place, is equal to the power of 10 by which the number can be divided.

The 2nd number to be the sum of 2 different squares: $10 = 1^2 + 3^2$.

10 is not, however, the difference of 2 squares, because it is of the form $4n + 2$. The sequence of numbers that are not the difference of 2 squares is 2 6 10 14 18 . . .

The 4th triangular number: $10 = 1 + 2 + 3 + 4$. There are 10 pins in a triangular array in a bowling alley. It is the only triangular number that is the sum of consecutive odd squares.

The 3rd tetrahedral number: $10 = 1 + 3 + 6$, where 1, 3 and 6 are the triangular numbers.

Among any 10 consecutive integers there is at least one that is relatively prime to all the others. [B. G. Eke]

$10! = 6!7!$ The only known solution to $n! = a!b!$ apart from the general pattern, $(n!)! = n!(n! - 1)!$. Also, $10! = 7!5!3!$

The base of Briggs's logarithms.

[57]

$n - \phi(n)$ is never equal to 10. The sequence of integers with this property continues: (10) 26 34 50 52 58 86 100 . . . [Guy 91]

The smallest number, apart from 1, which is not the sum of a square and a prime. The next two are 25 and 34. [Sloane 73]

Euler conjectured in 1782 that two mutually orthogonal Latin squares do not exist of order $4n + 2$. This is true for order 6, but false for orders 10, 14, . . . , as Bose, Shrikhande and Parker proved in 1959. In the figure, every bold digit appears once in each row and column, and so does every *italic* digit. Moreover, every pair of digits from 00 to 99 appears just once in the figure.

46	57	68	70	81	02	13	24	35	99
71	94	37	65	12	40	29	06	88	53
93	26	54	01	38	19	85	77	60	42
15	43	80	27	09	74	66	58	92	31
32	78	16	89	63	55	47	91	04	20
67	05	79	52	44	36	90	83	21	18
84	69	41	33	25	98	72	10	56	07
59	30	22	14	97	61	08	45	73	86
28	11	03	96	50	87	34	62	49	75
00	82	95	48	76	23	51	39	17	64

The news of Euler's failure, unlike most mathematical discoveries, made headlines in the newspapers, and Bose, Shrikhande and Parker were nicknamed 'Euler's spoilers'.

Desargues's theorem defines a configuration of 10 lines, with 3 points on each line and 3 lines passing through each point.

Take a number, and multiply its digits together. Repeat with the answer, and repeat again until a single digit is reached. The number of steps required is called the multiplicative persistence of the number.

10 is the smallest number with multiplicative persistence of 1. The smallest numbers with multiplicative persistence 2 to 8 are:

(1)	2	3	4	5	6	7	8
(10)	25	39	77	679	6788	68,889	2,677,889

The smallest number with multiplicative persistence of 11 is 277,777,788,888,899. No number less than 10^{50} has a greater multiplicative persistence and it is conjectured that there is an upper limit to the multiplicative persistence of any number. [Sloane, JRM v6]

The decimal system

The Greek philosopher Aristotle and the Roman poet Ovid agreed that we count in 10s because we have 10 fingers. It is as reasonable to conclude that some cultures count in 5s based on individual hands, and that counting in 20s is based on using the hands and the feet.

To count a small number of objects is not difficult. Indeed, it is sufficient to have a standard sequence of names for them, such as one–two–three–four–five–six–seven–eight–nine–ten. The difficulty arises when it is desired to count many objects. The necessarily limited set of basic names must somehow be repeated in different combinations. The clearer and simpler the system of repetitions, the easier it will be to count, and, just as significantly, to calculate.

The ancient Egyptians recorded numbers by grouping symbols for powers of 10. This is as cumbersome as the Roman system, still used occasionally in public inscriptions.

Our modern system of counting in 10s, and the variants that are used in computers, such as bases 2, 8, and 16 and alternatives that are sometimes proposed, such as the duodecimal system or base 12, are all founded on two principles, the use of zero and the place-value principle.

When the value of a numeral depends only on where in the number it appears, a limited set of numerals, only 0 and 1 in binary, can be used to count in a very simple and regular manner, as high as we please, and to calculate by simple and powerful algorithms, known to school pupils as 'sums' though they do much more than merely add numbers together.

Pierre Simon de Laplace remarked that this very simplicity 'is the reason for our not being sufficiently aware how much admiration it deserves'.

The Roman system used the leters I, V, X, L, C, D, M to stand for 1, 5, 10, 50, 100, 500, 1000. These numbers go up in jerks, alternately increasing fivefold and doubling.

The value of a digit in our system increases tenfold with every step to the left: 1 − 10 − 100 − 1000 − 10,000, and so on.

Unfortunately, 10 is not an ideal base for a system in which merchants and dealers have to measure small quantities, fractions of a whole, because only halves and fifths can be represented by whole numbers. Even a simple fraction like a quarter has to be represented by a fraction of 10ths. Consequently, although using a number system based on 10, an extraordinary variety of systems of weights and measures was used throughout Europe in historical times based on mixtures of units. They all agreed in using 8ths, 12ths, 20ths, 60ths, 24ths, anything but the awkward 10th.

Not until 1791 when the Paris Academy of Sciences recommended a new metric system did any generally acceptable and uniform system start to emerge. 1 metre was defined to be 1/40,000,000 part of a circumference of the earth through the poles. The ratios between units were to be always powers of 10. Greek and Latin prefixes were used for larger and smaller units, respectively.

The metre as the unit of length was used to define units of volume and mass, and today all scientific measurements are based on the metric system.

For mathematicians, on the other hand, 10ths posed no problem. All they wanted was a system for representing indefinitely small quantities that was as easy to use as the usual base 10 for whole numbers. Adam Riese took a large step forward in 1522 when he published a table of square roots, explaining that the numbers had been multiplied by 1,000,000 and so the roots were 1000 times too large.

François Viète, in 1579, published a book in which he used decimal fractions as a matter of course, and recommended their use to others, and Simon Stevin in 1585 published a 7-page pamphlet in which he explained decimal fractions and their use. Stevin also had the foresight to recommend that a decimal system should be used for weights and measures and coinage and for measuring angles.

There is a postscript to the history of decimal fractions. The notation of decimals still varies between the English, who place the decimal point at the middle level, the Americans who place it on the line, and continental Europe where a comma is used.

The Pythagoreans

Pythagoras and his disciples taught that everything is Number. Numbers to them meant strictly whole numbers, integers. Fractions were considered only as ratios between integers.

The Greeks distinguished between *logistike* (whence our term logistics), which meant numeration and computation, and *arithmetike*, which was the theory of numbers themselves. It was *arithmetike* that Plato, a convinced Pythagorean, insisted should be learned by every citizen of his ideal Republic, as a form of moral instruction. It was a profound shock to their philosophy when $\sqrt{2}$ was discovered to be not the ratio of 2 integers, although it was undoubtedly a length and therefore, to the Greeks who thought of numbers geometrically, a number or ratio of numbers.

Pythagoras himself or his disciples discovered that harmony in music corresponded to simple ratios in numbers. Indeed, it was this discovery that provided the earliest support for their doctrine. Aristotle records that, 'They supposed the elements of number to be the elements of all things, and the whole heaven to be a musical scale and a number.'

The octave corresponds to the ratio 2:1 because if the length of a musical string is halved, it sounds one octave higher. The ratio 3:2 corresponds to the fifth and 4:3 to the fourth. Somewhat less harmonious intervals were represented by rather larger numbers. A single tone was the difference between a 5th and a 4th, and was therefore 9:8, which is 3:2 divided by 4:3.

(The problem of constructing a complete scale is very complex, and has engaged the efforts of musicians to the present day. All solutions involve approximation. It is not possible for example for a fixed scale, such as a piano possesses, to include all the perfect 5ths and 4ths that the performer would like. The violinist has an advantage here over the pianist. The solution that divides the octave into 12 equal tones gets none of them perfectly correct.)

The basic ratios could be represented in the sequence 12:9:8:6 and the sum of these numbers, 35, was called harmony.

More commonly, the Pythagoreans thought of these ratios as involving only 1, 2, 3 and 4, whose sum is 10, which is the base of our counting system. How elegantly everything fits together! No wonder they felt confirmed in their diagnosis of the vital significance of Number.

The number 10 can also be represented as a triangle, which they called the *tetraktys*. To the Pythagoreans it was holy, so holy that they even swore oaths by it.

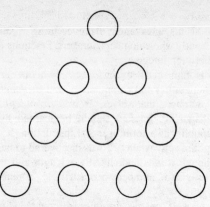

Later Pythagoreans described many other tetraktys. Magnitude, for example, comprised point, line, surface and solid.

The primitive aspects of Pythagorean belief died out very slowly. Their musical discoveries did not die at all. They were true science, two thousand years before modern science displayed the whole numbers in the chemist's Periodic Table or the physicist's model of the atom.

Precisely because music was for so long a unique example of genuine numbers-in-science, it had an overwhelming effect. Leibniz wrote, 'Music is a secret arithmetical exercise and the person who indulges in it does not realize that he is manipulating numbers.' That is not quite correct. Early classical composers, before the advent of Romanticism, were often quite deliberate in their use of mathematical patterns to structure their music.

Unlike the Greeks, we are not limited to the whole numbers, and today science often seems to be soaked in rational approximations, rational results from experimental observation. Yet underneath the complexity of modern science, the integers may still occupy a central role. Daniel Shanks gives many examples of their role in modern science. To relate just one of his examples, why is the force of gravity at double the distance reduced by a factor of 4? Why is the factor apparently 4 exactly, rather than 4 approximately? Probably because we live in a space of exactly 3 dimensions.

The Pythagoreans' faith in the whole numbers may be vindicated yet.

[Shanks, *Solved and Unsolved Problems in Number Theory*, Spartan Books, 1962]

11

The 5th prime.

The smallest repunit, apart from 1.

Because 11 = 10 + 1, there is a simple test for divisibility by 11. Add and subtract the digits alternately, from one end. (Either end may be chosen as the starting point.) If the answer is divisible by 11, so is the number. This is equivalent to adding the digits in the odd positions, and in the even positions, and subtracting one answer from the other.

11 appears as a factor, and a multiple, though not by itself, in the imperial system of measuring length. $5\frac{1}{2}$ yards was one rod, pole or perch; 22 yards is a chain; 220 yards a furlong; and $1760 = 11 \times 160$ yards makes 1 mile.

11 is the only palindromic prime with an even number of digits.

Given any 4 consecutive integers greater than 11, there is at least one of them that is divisible by a prime greater than 11.

The smallest number which is the sum of a square and prime in 3 ways. [Sloane 73]

The 7th Ulam number. The Ulam sequence starts 1 2 3, and each new term is the next to be uniquely the sum of 2 previous terms in the sequence. The sequence continues: (1 2 3) 4 6 8 11 13 16 18 26 . . . [Sloane 557]

The world we live in is apparently 3-dimensional, or 4-dimensional when time is counted as an extra dimension. According to the latest physical theory of supersymmetry, space is most easily described as 11-dimensional. Seven of the dimensions are 'curled up on themselves'. Their physical effects would be directly observable only on a still inaccessible scale billions of times smaller even than that of subatomic particles.

The Lucas numbers

11 is the 5th number in the Lucas sequence: 1   3   4   7   11   18   29   47   76   123   199   322 . . .

This sequence (which is sometimes started 2 1 3 4 7 . . .) is closely related to the Fibonacci sequence. Each term is the sum of the previous 2 terms, and the ratio of successive terms tends to the Golden Ratio as a limit. It is a curiosity that the Lucas sequence also has an easy-to-remember convergent to φ, 322/199.

There is a formula for the *n*th term that is very similar to the formula for the *n*th Fibonacci number:

$$L_n = \frac{(1 + \sqrt{5})^n}{2^n} + \frac{(1 - \sqrt{5})^n}{2^n}$$

or $L_n = a^n + b^n$ where a and b are the roots of $x^2 = x + 1$.

The two formulae share a useful property. The 2nd term decreases so rapidly that the Lucas numbers can be calculated by finding the nearest integer to the powers of φ: thus, $\phi^5 = 11 \cdot 09017$ and $L_5 = 11$.

Lucas discovered many properties of the Fibonacci sequence, and

[63]

studied general Fibonacci sequences, in which each term is the sum of the previous two terms, but the initial terms are not necessarily 1 and 1, or 1 and 3. He used the Fibonacci and the Lucas sequences to construct tests for the primality of the Mersenne numbers.

The Lucas numbers can be expressed as sums of Fibonacci numbers:

$$L_n = F_{n-1} + F_{n+1}$$

It is always true that F_n divides F_{mn}. For small values of m, the ratio is known and can be expressed in terms of Lucas numbers, for example:

$$F_{2n} = F_n L_n$$
$$F_{3n} = F_n(L_{2n} + (-1)^n)$$

Squaring the Fibonacci numbers, then alternately subtracting and adding 4, produces the squares of the Lucas numbers:

$$5 \times 1^2 - 4 = 1^2 \qquad 5 \times 1^2 + 4 = 3^2$$
$$5 \times 2^2 - 4 = 2^2 \qquad 5 \times 3^2 + 4 = 7^2 \qquad \text{and so on.}$$

Naturally there are many formulae connecting the Lucas numbers alone, for example, $L_{2n} = L_n^2 - 2(-1)^n$.

12

There are 12 months in the year, divided roughly into 4 seasons, 12 signs of the Zodiac, divided into 3 sets of 4 each, and 12 hours, repeated through each day and night.

There are 12 different pentominoes, if pieces can be flipped over. Otherwise there are 18.

12 is divisible by the sum of its digits and by their product.

The product of the proper divisors of 12 is $12^2 = 144$.

$12^2 = 144$ and, reversing all digits, $21^2 = 441$.

The same pattern fits $13^2 = 169$ and $31^2 = 961$ and other squares of numbers with sufficient small digits.

$12 = 3 \times 4$, trivial in itself, but curious when continued with $56 = 7 \times 8$.

$12\sigma(12) = 14\sigma(14)$.

There are 12 ways of arranging 8 queens on a standard chessboard (ignoring reflections and rotations) so that no queen attacks any other. If reflections and rotations are counted separately, there are 92.

It takes 12 knights to either attack or occupy every square of the standard chessboard. Their positions are c2, c3, d3 and so on, with rotational symmetry. [Gardner, *Mathematical Magic Show*, 1978, 194]

There are 12 tones in the modern 12-tone musical scale.

12 identical spheres can touch one other such sphere, each of the outer

spheres touching the central sphere and 4 others. The numbers of spheres that can touch one sphere in higher dimensions up to dimension 9 are:

dim.4	dim.5	dim.6	dim.7	dim.8	dim.9
24	40	72	126	240	272

The dual polyhedra, the cube and the octahedron, each have 12 edges.

Abundant numbers

12 is the first abundant number, meaning that it is less than the sum of its factors excluding itself: $1 + 2 + 3 + 4 + 6 = 16$.

There are only 21 abundant numbers not greater than 100, starting 12, 18, 20, 24, 30, 36 . . .

They are all even. Abundant numbers are essentially numbers with enough different prime factors. Most numbers have very few factors, and are deficient, that is, they are greater than the sum of their factors. All primes and powers of primes are deficient. The least deficient prime powers, as it were, are the powers of 2. The divisors of 2^n excluding itself sum to $2^n - 1$, only 1 less than the original number. For this reason such numbers are sometimes called almost perfect.

Dividing the abundant from the deficient numbers are the very rare perfect numbers, exactly equal to the sum of their divisors.

All multiples of a perfect or abundant number are also abundant. Any divisor of a perfect or deficient number is also deficient.

$\sigma(n)$ denotes the sum of the divisors of n, including n; $\sigma(12)/12 = 12 + 16/12 = 28/12$ or $7/3$, which is a record for numbers up to 12. Any number that sets a record for $\sigma(n)/n$ is called superabundant. It is known that there are an infinite number of superabundant numbers.

The duodecimal system

Although we take counting in 10s for granted, there are disadvantages in using 10 as a base or as a ratio between standard measures. It is especially annoying that a simple fraction like a third cannot be represented exactly, but only as a repeating decimal fraction.

The duodecimal system, based on 12, allows thirds, quarters and sixths to be expressed very simply. The 12 months of the year divide naturally into 4 seasons of 3 months each, the 12 signs of the Zodiac divide into 4 groups of signs associated with fire, air, earth and water respectively. In many calendars the 12 months are divided into 6 short months and 6 long months.

It is also as easy to test a number for divisibility by 2, 3, 4, 6, 8, 12, 16, 24 in base 12 as it is to test for divisibility in base 10 by 2, 5, 10, 20.

These were important advantages when calculation itself was a subtle art and difficult to learn, so important that all over Europe the 10 system, based on our 10 fingers, was mixed up with systems of units based on ratios of 2, 4, and especially 12, or combinations of 10 and 12.

Plato, describing his ideal state, established its coinage and weights and measures, the voting districts and representation in the assembly, and even the fines to be levied for offences, on a duodecimal system.

The Romans used only duodecimal fractions. When Pliny the Elder estimated the area of Europe to the whole world he stated that it was 'somewhat more than the third and the eighth part of the whole earth' using Roman fractions in the Egyptian manner, instead of saying 'eleven twenty-fourths'. [Menninger] They called one twelfth *uncia*, whence our word *ounce*. When an uncia was not small enough, it was divided into 24 scruples, which might be subdivided again. The smallest unit, 1 calcus = 1/8 scruple = 1/192 uncia = 1/2304 unit.

Elsewhere the sexagesimal system, based on 60, has been used, especially for scientific calculation. Because $60 = 5 \times 12$ it has the advantages of 10 and 12 combined.

We still count 12 inches to the foot, as well as using the metric system. Everyone was familiar with the 12 pence in 1 shilling before decimalization in 1971. This originated in Charlemagne's monetary standard: 1 libra = 20 solidi = 240 denarii, whence our '£' sign and 'd' for pence.

We still talk of a dozen or dozens, though it is coming to mean 'quite a large number' rather than any exact figure, and the gross or dozen dozen is almost obsolete.

In England there used to be a long hundred of 120 units and a short hundred of 100 units. It was often necessary to state whether 'one hundred' was by the 12-count or the 10-count. The great hundred of 120 units is still used in Germany and Scandinavia.

Buffon proposed that a duodecimal system be universally adopted, for counting and for all measures and coinage. So did Isaac Pitman, the inventor of Pitman shorthand, Herbert Spencer, the philosopher, H. G. Wells and Bernard Shaw, and many others.

In 1944 The Duodecimal Society was established as a voluntary, non-profit-making organization in New York State. Its aims were 'to conduct research and education of the public in mathematical science, with particular relation to the use of Base Twelve in numeration, mathematics, weights and measures, and other branches of pure and applied science'. The Duodecimal Society proposed to add the letter X to represent 10 and E to represent 11, and claimed that counting by dozens can be learned by anyone in about half an hour. They were soon arguing that the terms decimal and decimal point were 'definitely improper' when referring to bases other than 10, as was reference to decimal fractions. Despite their enthusiasm there is no chance at all of the change to duodecimal ever being made. Indeed, over the last century or so the change has gone the other way, ever since the metric system was introduced.

The dodecahedron

The number of faces of a dodecahedron, the 4th of the Platonic solids, is 12. It also has 20 vertices and 30 edges, and is the dual of the icosahedron. If the mid-points of neighbouring faces of a regular dodecahedron are joined, for example, they form a regular icosahedron.

The regular icosahedron can be seen as an antiprism with pentagonal ends, plus 2 pentagonal pyramids. Not surprisingly, the presence of regular pentagons means the presence also of the Golden Section. In particular, if opposite edges of the antiprism are joined, then 3 rectangles, whose sides are in the Golden Ratio, are obtained, at right-angles to each other.

It is an extraordinary fact, which at first seems absurd, that if a dodecahedron and an icosahedron are each inscribed in identical spheres, the dodecahedron occupies a greater volume, although the icosahedron has more faces and would seem therefore naturally to 'fit better'. In fact the dodecahedron occupies approximately 66·5% of the sphere, the icosahedron only 60·56%.

The rhombic dodecahedron, first described by Kepler, also has 12 faces. Imagine cubes packed together to fill space. The 6 cubes adjacent to any one cube can each be cut into 6 pyramids by joining their centres to the vertices. If these pyramids are then glued to their facing cubes, each cube becomes a rhombic dodecahedron, and the rhombic dodecahedrons pack the space completely, just as the cubes did, with the difference that each rhombic dodecahedron has double the volume of the corresponding cube.

13

A notoriously unlucky number. This superstition has been linked to the 13 who sat at table at the Last Supper, but it probably originated only in the medieval period. There is a word for fear of the number 13, such as fear of living on the 13th floor of an apartment block: triakaidekaphobia, from the Greek for 'fear of thirteen'.

There are 13 times 4 weeks in a year, and 13 cards in each suit of a standard pack.

Ironically, 13 is the 5th Lucky number, and also the 6th prime and the 7th Fibonacci number.

13 is the second smallest prime, p, the period of whose reciprocal is $\frac{1}{2}(p - 1)$: $1/13 = 0·076923\ 076923 \ldots$ (1/3 is the smallest such prime).

Exactly half the multiples of 1/13 from 1/13 to 12/13 have periods that are a cyclic permutation of this string. The other multiples all have periods that are cyclic permutations of 153846.

The sequence of digits forms a pattern that is more apparent when arranged as in this figure:

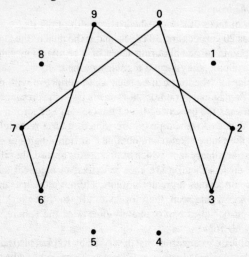

12! + 1 is divisible by 13^2.

A torus can be sliced into 13 parts with just 3 plane cuts.

The smallest Emirp: a prime which is a different prime when reversed. The sequence continues: 17 31 37 71 73 79 97 107 113 . . .

The Pellian equation $x^2 - ny^2 = 1$ is famous for having surprisingly large minimum solutions for certain values of n. $n = 13$ is the first of these, for which $x = 649$, $y = 180$ is the smallest solution. The next is $n = 29$, with minimum solution, $x = 9801$, $y = 1820$.

The Archimedean polyhedra

There are 13 Archimedean polyhedra, named after Archimedes who wrote a book on them, now lost. Kepler was the first modern mathematician to describe them. They are described as semi-regular, because their edges and vertices are all the same, and their faces are all regular polygons, though not all of the same type. Two infinite classes of polyhedra are also semi-regular, the regular prisms and the regular antiprisms. Kepler also discovered the smaller and greater stellated dodecahedrons, re-discovered with two other polyhedra that are regular but not convex by Poinsot.

There are also 13 dual Archimedean polyhedra, whose vertices, but not the faces themselves, are regular, and a number of stellations of the Archimedean solids, corresponding to the Kepler–Poinsot stellations of the Platonic solids.

There are also a number of beautiful compound polyhedra, which demonstrate the symmetry of the vertices of the inscribed solid.

What convex polyhedra are possible if all symmetry conditions are dropped, except for the regularity of the faces? This was answered only recently, in the 1960s. The regular-faced convex polyhedra are: the regular prisms and antiprisms, the 5 Platonic solids, the 13 Archimedean polyhedra, and 92 others. [Johnson, *Canadian Journal of Mathematics* v18]

The theorem of Pythagoras and Pythagorean triples

The theorem of Pythagoras, that in a right-angled triangle the sum of the squares on the shorter sides is equal to the square of the hypotenuse, has been familiar to generations of schoolchildren. Indeed, it is so famous that it is even the punch line of a joke, '. . . which proves that the squaw on the hippopotamus is equal to the sum of the squaws on the other two hides.'

More proofs have been published of Pythagoras's theorem than of any other proposition in mathematics, several hundred in all.

The $3-4-5$ triangle is the simplest example of a Pythagorean triangle, that is, a right-angled triangle with integral sides, but it is only one of an infinite set, which continues with $5-12-13$, hence the present entry, $6-8-10$ which is not *primitive* because it is just a multiple of the $3-4-5$ triangle, and then $7-24-25$.

The Babylonians about 2000 BC were familiar with Pythagorean triangles, though we do not know what they called them. The famous cuneiform tablet, Plimpton 322, lists 15 sets of numbers that are the sides of right-angled triangles.

The author of this tablet apparently knew that the numbers $2pq, p^2 - q^2$ and $p^2 + q^2$ are the sides of a right-angled triangle. (It is also true that the sides of any right-angled triangle that do not have any common factor are of this form.)

The Greeks almost certainly obtained at least the idea from further east, and either Pythagoras himself or one of his disciples discovered a proof of the geometrical proposition.

The $3-4-5$ triangle has a number of properties not shared by other Pythagorean triangles (apart from multiples such as $6-8-10$).

It is the only Pythagorean triangle whose sides are in arithmetic progression. It is also the only triangle of any shape with integral sides, the sum of whose sides (12) is equal to double its area (6).

Curiously, there is at least one other Pythagorean triangle whose area is expressed with a single digit: the triangle 693–1924–2045 has area 666,666. On average, one-sixth of all Pythagorean triangles have areas ending in the digit 6, in base 10; one-sixth end in 4 and the other two-thirds end in 0. [W. P. Whitlock Jnr]

There is an infinite set of triangles such that the hypotenuse and one leg differ by 1. They follow this pattern:

$$3^2 = 9 = 4 + 5 \qquad 3^2 + 4^2 = 5^2$$
$$5^2 = 25 = 12 + 13 \qquad 5^2 + 12^2 = 13^2$$
$$7^2 = 49 = 24 + 25 \qquad 7^2 + 24^2 = 25^2 \ldots$$

There is also an infinite number of triangles whose legs differ by 1, though they are not so simple to calculate. Starting with the formula for the sides: $2pq$, $p^2 - q^2$ and $p^2 + q^2$, where p and q are any 2 integers, if p and q generate a triangle whose legs differ by 1, the next such triangle is generated by q and $p + 2q$.

The 3–4–5 triangle is generated from the formula by 1 and 2, so the next almost-isosceles triangle will be generated by 2 and 5. It is 20–21–29. By applying the rule $(p, q) \rightarrow (q, p + 2q)$ repeatedly, we obtain this sequence: 1 2 5 12 29 70 169 408 ... Taking any two successive members of the sequence for generators produces an almost isosceles Pythagorean triangle. Of course, the triangle can never be actually isosceles, because $\sqrt{2}$ is irrational. The same sequence of numbers occurs in the best approximations to $\sqrt{2}$ by fractions.

The formula already given for the sides of a right-angled triangle implies that the length of the hypotenuse is also the sum of 2 squares. Girard knew and Fermat a few years later proved the beautiful theorem that every prime of the form $4n + 1$, that is, the primes 5, 13, 17, 29, 37, 41, 53 ..., are each the sum of 2 squares in exactly one way. Primes of the form $4n + 3$, such as 3, 7, 11, 19, 23, 31, 43, 47 ..., are never the sum of 2 squares.

Leonardo of Pisa already knew that the product of 2 numbers that are each the sums of 2 squares is also the sum of 2 squares. It follows that the square of any of these numbers, say 13^2, is the sum of 2 squares, and therefore the hypotenuse of a right-angled triangle. The converse however is more complicated; thus, $17^2 + 144^2 = 145^2$ and 145 is not prime, though it is the product of 5 and 29 both of which are primes of the form $4n + 1$.

The square of the hypotenuse of a right-angled triangle is also the difference of 2 cubes; thus, $13^2 = 8^3 - 7^3$.

There are other ways to obtain Pythagorean triples. Take any pair of consecutive odd or even numbers, and add their reciprocals. For example, $1/3 + 1/5 = 8/15$. Then 8 and 15 are the legs of a right-angled triangle: in fact, $8^2 + 15^2 = 17^2$. This method is equivalent to making one of the generators in the usual formula equal to 1, so produces only a subset of all possible triangles.

If any 2 of the sides of a right-angled triangle are taken as generators for a new triangle, then the resulting triangle will contain the square of the 3rd side of the original triangle as one of its sides. [W. P. Whitlock Jnr] Thus, take 3 and 4 from the 3−4−5 triangle. The new triangle is 7−24−25, which contains 5^2 as one of its sides.

14

In the imperial system of weights and measures, the number of pounds weight in 1 stone.

Also the number of days in a fortnight.

14 is the 3rd square pyramidal number: $14 = 1 + 4 + 9$.

$\sigma(14) = \sigma(15)$, the first pair satisfying $\sigma(n) = \sigma(n + 1)$. The sequence continues: (14) 206 957 1334 1364 1634 2685 2974 ... [Guy 68]

14 is the smallest even number, n, such that $\phi(m) = n$ has no solution. The sequence of such numbers continues: (14) 26 34 38 50 62 68 74 76 ...

Equilateral triangles with integral sides, which have irrational areas, can be approximated by Heronian triangles with integral sides and area. The 1st approximation is the Pythagorean triangle with sides 3, 4 and 5, and area 6. The 2nd approximation is 13, 14, 15 with area 84 (twice its perimeter), where 14 is calculated as $4^2 - 2$. The 3rd approximation is 193, 194, 195, where $194 = 14^2 - 2$, and the 4th is 37,633−4−5, and so on.

The smallest repfigit number, meaning that if a generalized Fibonacci sequence is started with these digits in sequence, each subsequent term being the sum of the previous 2, then the sequence, which starts 1 4 5 9 14 ..., includes 14. The sequence of repfigit numbers starts: 14 19 28 47 61 75 197 742 1104 ... [JRM v26 195]

There are 14 distinct ways to represent 1 as the sum of 4 unit fractions. The totals for 5 and 6 unit fractions are 147 and 3462 respectively. [David Singmaster: Guy 162]

14·13472 5 ...

According to the Riemann Hypothesis, all the non-trivial zeros of the complex-valued Riemann zeta function, $\zeta(z)$, are of the form $1/2 + t\sqrt{-1}$, where $t > 0$. The first zero is $1/2 + i(14·13472\ 5 ...)$.

15

The first product of 2 odd primes.

The sum of the rows, columns and diagonals of the smallest magic square.

The equation $x^2 + 7 = 2^n$ has solutions for only 5 values of n: 3, 4, 5, 7 and 15. For $n = 15$, the solution is $x = 181$. [AMM v94 59]

If n is greater than 15, then there is at least one number between n and $2n$ which is the product of 3 different primes. [Sierpinski, *250 Problems in Elementary Number Theory*, no. 94]

Triangular numbers

15 is the 5th triangular number. There are 15 balls in a snooker triangle.

The Greeks named the triangular numbers, and formed them by adding up the series $1 + 2 + 3 + 4 + 5 \ldots$

The general formula for the nth triangular number, denoted by T_n, is $\frac{1}{2}n(n + 1)$ and the sequence starts: 1 3 6 10 15 21 28 ...

$\frac{1}{2}n(n + 1)$ is also a binomial coefficient, so the triangular numbers should appear in Pascal's triangle. They do, as the 3rd diagonal in each direction.

The triangular numbers are the simplest of the polygonal numbers. There are many relationships between them. Each square number is the sum of 2 successive triangular numbers. Alternatively, as Diophantus knew, each odd square is 8 times a triangular number, plus 1.

Each pentagonal number is the sum of 3 triangular numbers in an especially simple way.

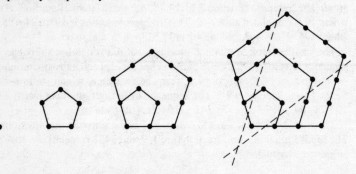

For every triangular number, T_n, there are an infinite number of other triangular numbers, T_m, such that $T_n T_m$ is a square. For example, $T_3 \times T_{24} = 30^2$.

On the other hand, the square of any odd number is the difference between two relatively prime triangular numbers.

Another relationship between triangular numbers and squares:

$$T_n = n^2 - (n - 1)^2 + (n - 2)^2 - (n - 3)^2 + (n - 4)^2 - \ldots \pm 1$$

There is a beautiful relationship between the triangular numbers and the cubes: $T_{n+1}^2 - T_n^2 = (n + 1)^3$, from which it follows that the sum of the first n cubes is the square of the nth triangular number, for example: $1 + 8 + 27 + 64 = 100 = 10^2$.

This points to a connection with the sums of 5th powers, because it is always true that $1^3 + 2^3 + 3^3 + \ldots + n^3$ divides $3(1^5 + 2^5 + 3^5 + \ldots + n^5)$. In fact, $\sum_1^n m^5 = \frac{1}{3}T_n^2(4T_n - 1)$. M. N. Khatri points out that adding the triangular numbers themselves produces this curious pattern:

$$T_1 + T_2 + T_3 = T_4$$
$$T_5 + T_6 + T_7 + T_8 = T_9 + T_{10}$$
$$T_{11} + T_{12} + T_{13} + T_{14} + T_{15} = T_{16} + T_{17} + T_{18}$$

and so on, from which he deduces among other facts that every 4th power is the sum of two triangular numbers. For example, $7^4 = T_{41} + T_{55}$.

Two relations between the triangular numbers alone: $T_n^2 = T_n + T_{n-1}T_{n+1}$, and $2T_nT_{n-1} = T_{n^2-1}$.

The series formed by summing the reciprocals of the triangular numbers converges: $1 + 1/3 + 1/6 + 1/10 + 1/15 + 1/21 + 1/28 + \ldots = 2$.

15 and 21 are the smallest pair of triangular numbers whose sum and difference (6 and 36) are also triangular. The next such pairs are 780 and 990, and 1,747,515 and 2,185,095. [Dickson] It happens that 6 is 'the only triangular number besides unity with fewer than 660 digits whose square is a triangular number'. [Beiler]

Some numbers are simultaneously triangular and square. The first is, of course, 1. The next 4 are 36, 1225, 41,616 and 1,413,721. The roots of these numbers, 1, 6, 35, 204, 1189 . . . , follow a simple pattern illustrated by $1189 = (204 \times 6) - 35$. These are found by using a fact already mentioned, that $8T_n + 1$ is always a square. If the triangular number is itself a square, say x^2, then we have the Pell equation: $8x^2 + 1 = y^2$.

The general formula is $\frac{1}{32}[(17 + 12\sqrt{2})^n + (17 - 12\sqrt{2})^n - 2]$.

There is also a rule for obtaining one solution from another: if T_n is a perfect square, then so is $T_{4n(n+1)}$.

On the other hand, no triangular number can be a cube, or 4th or 5th power.

Charles Trigg gives examples of palindromic triangular numbers. The smallest, apart from 1, 3 and 6, are 55, 66, 171, 595, 666 and 3003. $T_{2662} = 3,544,453$, so the number itself and its index, 2662, are both palindromic. T_{1111} and $T_{111,111}$ are 617716 and 6172882716 respectively.

16

The 4th square and second 4th power, after 1.

The first square to be the sum of 2 triangular numbers in 2 ways: $16 = 6 + 10 = 1 + 15$.

All sufficiently large numbers are the sum of at most 16 4th powers.

$16! = 14!5!2!$

$\sigma(16) = 31$ is prime. The sequence of n for which $\sigma(n)$ is prime runs: 2 4 9 16 25 289 ...

Euler showed that the only integer solution to $a^b = b^a$ is $4^2 = 2^4 = 16$.

The Pythagoreans knew that 16 is the only number that is the perimeter and the area of the same square.

16, like 12, has often been proposed as a base for a new system of counting. J. W. Mystrom in the nineteenth century proposed that the numbers 1 to 16 in this system should be named: *an*, *de*, *ti*, *go*, *su*, *by*, *ra*, *me*, *ni*, *ko*, *hu*, *vy*, *la*, *po*, *fy* and *ton*.

Order-4 magic squares
The first 16 numbers can be arranged in many ways to make an order-4 magic square in which each row and column and both the diagonals have the same sum, which will always be 34.

The illustration shows the magic square from Dürer's engraving *Melencolia*. The numbers in the middle of the bottom row give the year in which it was made, 1514.

Many magic squares, like the 3 × 3, have extra and elegant properties. This one was described by Alfred Moessner:

12	13	1	8
6	3	15	10
7	2	14	11
9	16	4	5

The sums of the cubes of the numbers along each diagonal are equal, to $4624 = 68^2$. The sums of the squares of the numbers in the 1st and 4th rows are equal. The same property is shared by the 2nd and 3rd rows, and by the 1st and 4th columns and the 2nd and 3rd columns. [Moessner, *Scripta Mathematica* v13]

The hexadecimal system

The base of the hexadecimal system, used in computers. To the usual numerals 0 to 9, the six letters A, B, C, D, E and F are added, standing for the numbers 10 to 15. Numbers are then constructed on the usual principles. Thus 6C5 stands for 5 units, C = 12 sixteens, and 6 sixteen-squareds, or $5 + 12 \times 16 + 6 \times 256 = 1733$.

Almost perfect numbers

16 is almost perfect, because its factors, excluding itself, sum to one less than itself: $1 + 2 + 4 + 8 = 15$.

All powers of 2 are almost perfect. Whether an odd almost perfect number exists is, of course, unknown. I say 'of course' because the existence of almost any kind of perfection in an odd number is 'not known'.

If a number's factors, excluding the number itself, sum to one more than the number, then the number is called quasi-perfect. It is known that a quasi-perfect number must be the square of an odd number, which is odd, but no one knows if any quasi-perfect numbers exist, which is odder. (*See 28* and *perfect numbers*.) [Guy] If a quasi-perfect does exist it is large, greater than 10^{35}, and has at least 7 distinct prime factors.

17

The 3rd Fermat prime: $17 = 2^{2^2} + 1$.

Gauss proved at the age of 18 that a regular polygon with a prime number of sides can be constructed with the use only of a straight edge and compasses only if the number of sides is of the form $2^{2^n} + 1$. It is possible therefore to construct a regular 17-gon with ruler and compasses only.

The period of $1/17 = 0\cdot05882\ 35294\ 11764\ 7$ is of maximal length, 16.

17 is the first sum of two distinct 4th powers: $17 = 1^4 + 2^4$.

17 is equal to the sum of the digits of its cube, 4913. It is the only prime to have this property. The only other such numbers are 1, 8, 18, 26 and 27, of which three are themselves cubes. [Zerger, JRM 25 248]

Choose numbers $a, b, c \ldots$ in the interval $(0,1)$ so that a and b are in different halves of the interval, a, b and c are in different thirds, a, b, c and d are in different quarters and so on. Not more than 17 such numbers can be chosen.

There are 17 essentially different symmetry patterns for a wallpaper design.

17 is the highest number whose square root was proved irrational by Theodorus.

According to Plutarch, 'The Pythagoreans also have a horror of the number 17. For 17 lies halfway between 16 ... and 18 ... these two being the only two numbers representing areas for which the perimeter (of the rectangle) equals the area.' [van der Waerden, *Science Awakening*, 1971]

$n^2 + n + 17$ is one of the best known polynomial expressions for primes. Its values for $n = 0$ to 15 are all prime, starting with 17 and ending with 257.

The smallest odd number which cannot be represented as the sum of a prime and twice a square. [Stern MM v66 45]

Every number greater than 17 is the sum of 3 integers greater than 1 which are relatively prime in pairs. [Sierpinski, *250 Problems in Elementary Number Theory*, no. 48]

The only known prime values for which $p^p - 1$ and $q^q - 1$ have a common factor less than 400,000 are 17 and 3313. The common factor is 112,643. [Stephens, MOC v25]

18

$18 = 9 + 9$ and its reversal, $81 = 9 \times 9$.

The pattern continues: $891 = 9 \times 99$; $198 = 99 + 99 \ldots$

This pattern works in any base. For example, in base 8: $7 + 7 = 16$ and $7 \times 7 = 61$. [Hsu, JRM v10]

The cube and 4th powers of 18 use all the digits 0 to 9 once each: $18^3 = 5832$ and $18^4 = 104,976$.

Equal to the sum of the digits of its cube, its 6th power and its 7th power: $18^3 = 5832$; $18^6 = 34,012,224$; $18^7 = 612,220,032$. [Stajsczak]

The largest of 4 integers n, for which $\phi(n) = 6$. The others are 7, 9 and 14.

18 is the smallest number that is twice the sum of its digits.

19

The 3rd number whose decimal reciprocal is of maximum length, in this case 18: $1/19 = 0 \cdot 05263\ 15789\ 47368\ 421$.

There is a simple test for divisibility by 19. $100a + b$ is divisible by 19 if and only if $a + 4b$ is.

19 is the 3rd centred hexagonal number: $19 = 1 + 6 + 12$.

There is only one way in which consecutive integers can be fitted into a magical hexagonal array, that is, so that their sums in all 3 directions are all equal. The numbers 1 to 19 can be so arranged.

The common sum of the unique magic hexagon is 38. As in the standard 3×3 magic square, the central cell is occupied by 5. [MG v71 217]

$19! - 18! + 17! - 16! + \ldots + 1$ is prime. The only other known numbers with this property are: 3, 4, 5, 6, 7, 8, 10, 15, 41, 59, 61, 105 and 160. [Guy 100]

$(19^{19} - 1)/(19 - 1)$ is prime. Other primes with this property are 2, 3, 7 and 31. [Guy 10]

12 trees can be planted in an orchard to create 19 rows of 3 trees each. It is conjectured that the maximum number of 3-rows is also 19 for 13 trees. [Sloane 982]

All integers are the sum of at most 19 4th powers.

20

The sum of the first 4 triangular numbers, and therefore the 4th tetrahedral number: $20 = 1 + 3 + 6 + 10$.

An icosahedron has 20 faces and its dual, the dodecahedron, has 20 vertices.

20 is the 2nd semi-perfect or pseudonymously pseudoperfect number, because it is the sum of some of its own factors: $20 = 10 + 5 + 4 + 1$. The smallest semi-perfect number is 12, which is also the 1st abundant number. The next are 18, 20, 24 and 30.

One leg of the second smallest right-angled triangle which is almost isosceles. The other leg is 21 and the hypotenuse 29. The smallest is the 3, 4, 5 triangle and the next three are: 119, 120, 169; 696, 697, 985; 4059, 4060, 5741. [Beiler]

The vigesimal system

20 has a special significance in many systems of counting and of weights and measures. Base 20, called vigesimal, was used by the Mayan astronomers and calendar makers whose culture flourished from the fourth century AD. Their system was positional and included a zero, centuries before the appearance of Indian numerals in Europe.

20 occurs in the old English coinage in '20 shillings in the pound' and in the imperial system of weights and measures. 20 is a score, and ages in biblical language are often expressed in scores: 'The days of our years are threescore and ten; and if by reason of strength they be fourscore years, yet is their strength labour and sorrow.' 'A score' or 'scores' survives as an expression for a largish number.

21

The 6th triangular number, and therefore the total number of pips on a normal dice.

If a square ends in the pattern $xyxyxyxyxy$, then xy is either 21, 29, 61, 69 or 84. The smallest example is: $508,853,989^2 = 258,932,382,121,212,121$. [Hunter, JRM v6]

21 is the smallest number of distinct squares into which a square can be dissected. The side of the dissected square is 112. [Duijvestijn, *Journal of Combinational Theory* v25]

21 is the smallest number which can be represented as the sum of at most 3 triangular numbers in 4 ways. [MOC v2 301]

The unique projective plane of order 4 has 21 points and 21 lines, with 5 points on each line and 5 lines through each point.

22

For $n = 22$, 23 and 24 only, the number of digits in $n!$ is equal to n.

The maximum number of pieces into which a pancake can be cut with 6 slices (see opposite). The sequence, starting with 1 slice, goes: 2 4 7 11 16 22 29 37 ...

22 is a palindrome, whose square is palindromic: $22^2 = 484$.

Many palindromes with sufficiently small digits have this property, for example, 11, 111, 1111, 121, 212 and so on.

$\sigma(22) = 36$, a square. The sequence of n for which $\sigma(n)$ is a square runs: 3 22 66 70 81 94 115 ...

Pentagonal numbers

22 is the 4th pentagonal number. The pentagonal numbers form the series: 1 5 12 22 35 51 70 ... The formula for the nth pentagonal number is $\frac{1}{2}n(3n - 1)$.

They can be formed in the Pythagorean manner as patterns of dots, forming successively larger pentagons (see p. 72). The formula, of course, produces values when *n* is a negative integer, in a way that the diagrams do not, so the sequence open in both directions reads: ... 40 26 15 7 2 0 1 5 12 22 35 51 60...

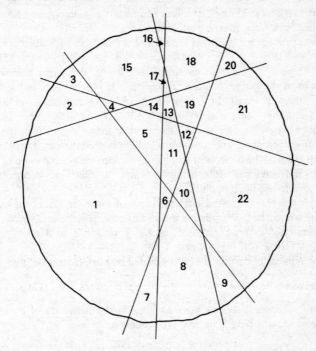

If these numbers are arranged in ascending order, a different pattern may be seen in their differences:

1 2 5 7 12 15 22 26 35 40 51 57 70 77 92 100
  1 3 2 5 3 7 4 9 5 11 6 13 7 15 8

The alternate differences form the natural numbers, 1, 2, 3, 4, 5, 6, 7 ... and the odd numbers, 1, 3, 5, 7, 9, 11, 13 ...

A very beautiful and important theorem was discovered by Euler, which involves the complete sequence in a surprising way. He started to multiply out the infinite product:

$$(1 - x)(1 - x^2)(1 - x^3)(1 - x^4) \ldots$$

and discovered that the first few terms were:

$$1 - x - x^2 + x^5 + x^7 - x^{12} - x^{15} + \ldots$$

At first he felt unable to prove, except by this informal induction, that the indices of the powers of x were indeed the pentagonal numbers, though the pattern was so strong that it was completely convincing, and he was satisfied to develop from it another theorem.

Euler proved that if $\sigma(n)$ is the sum of the divisors of n, then $\sigma(n) = \sigma(n - 1) + \sigma(n - 2) - \sigma(n - 5) - \sigma(n - 7) + \sigma(n - 12) + \sigma(n - 15) - \sigma(n - 22) - \ldots$ The sum continues as long as the terms represent the sum of the factors of positive numbers. If $\sigma(0)$ appears as the last term, then it must be replaced by n. For example, $\sigma(12) = \sigma(11) + \sigma(10) - \sigma(7) - \sigma(5) + \sigma(0) = 12 + 18 - 8 - 6 + 12 = 28$.

This relationship can be used to calculate the values of $\sigma(n)$ if you know the appropriate previous values, which is itself curious, because to find the sum of the divisors of a number you apparently need to know its factors and therefore whether or not it is prime, but none of this information is needed to use the formula!

Euler was also interested in the partitions of a number, that is, the ways in which it can be represented as the sum of other positive integers. 5 can be partitioned in 7 ways: 5 is 5, or $4 + 1$, or $3 + 2$, or $3 + 1 + 1$ or $2 + 2 + 1$ or $2 + 1 + 1 + 1$ or $1 + 1 + 1 + 1 + 1$.

The number of partitions of n is denoted by $p(n)$. It turns out that:

$$p(n) = p(n - 1) + p(n - 2) - p(n - 5) - p(n - 7) + p(n - 12) + \ldots$$

This sequence also continues as long as the partitions of positive numbers are involved. This time $p(0)$ counts as 1.

22·45915 77183 61045 47342 7152 . . .
π^e

It is not known whether this number is rational or irrational.

23
23 and 29 are the 1st pair of consecutive primes differing by 6.

23 is one of only 2 integers that actually needs 9 cubes to represent it. The other is 239. $23 = 2 \times 2^3 + 7 \times 1^3$.

It is required, of course, that the cubes be positive. If negative cubes are allowed then, for example, 23 is equal to $3 \times 2^3 + (-1)^3$, a total of only 4 cubes. In general, if negative cubes are allowed, then it is definitely known that all integers that do not leave a remainder of 4 or 5 on division by 9 can be represented as the sum of only 4 cubes.

23 is the 4th prime the period of whose reciprocal is of maximum length.

The smallest number of rigid rods of unit length required to brace a square is 23.

23 is the largest integer that is not the sum of distinct powers.

23! is 23 digits long.

If there are 23 or more people in a room the probability that at least 2 of them have the same birthday is greater than 50:50. If there are at least 88 people in the room, there is a better than 50:50 chance that 3 of them have the same birthday. If the dates of birthdays are not randomly scattered through the year, then these chances increase. [MG v70 228]

The smallest number which is not the sum of two Ulam numbers. [Guy 109]

For every n greater than 23, none of the binomial coefficients $\binom{n}{k}$ are square-free.

23·10345 . . .

The sum of the reciprocals, $1 + 1/2 + 1/3 + 1/4 + 1/5 + \ldots$ is unbounded. By taking sufficiently many terms, it can be made as large as one pleases. However, if the reciprocals of all numbers that when written in base 10 contain at least one 0 are omitted, then the sum has this limit, 23·10345 . . . [Boas and Wrench, AMM v78]

23·14069 26327 79269 00572 9086 . . .

e^{π}

This number is transcendental. Note that $e^{\pi} - \pi = 19\cdot99909\ 998 \ldots$

24

The number of hours in a day. Also 24 scruples in an ounce, and 24 grains in a pennyweight.

24 is divisible by the sum of its digits and by their product.

The smallest composite number, the product of whose proper divisors is a cube. $2 \times 3 \times 4 \times 6 \times 8 \times 12 = 24^3$.

The sum of the first 24 squares, which is the 24th square pyramidal number, is itself a square:

$$1^2 + 2^2 + 3^2 + \ldots + 24^2 = 70^2$$

This is the only solution to this pattern, though other sequences of consecutive squares not starting with 1 can sum to a square. For example, $18^2 + 19^2 + \ldots + 28^2 = 77^2$.

The sum of the squares of the divisors of 24 equals the sum of the squares of the divisors of 26. [Guy 68]

The smallest possible integral area of a scalene, obtuse-angled triangle with integral sides, 4, 13 and 15.

The smallest number representable as the sum of distinct Fibonacci numbers in 5 ways. [FQ v4 305]

Divisible by sum and product of its digits: the sequence of such numbers goes: 1   2   3   4   5   6   7   8   9   12   24   36   111   112   132   135 . . . [MM v63 10]

The smallest number which is the value of $\sigma(n)$ for three values of n: 14, 15 and 23. The next such numbers are: 42, 48, 60, 84, 90, . . .

If identical spheres in space of 24 dimensions are arranged in a Leech lattice, each sphere will touch 196,560 other spheres. This is almost certainly the densest possible sphere-packing in 24 dimensions. Suitable cross-sections of the Leech lattice packing give rise to the densest known packings in all lower dimensions, except for dimensions 10, 11 and 13. [Sloane, 'The Packing of Spheres', *Scientific American*, Jan. 1984]

Factorials
$24 = 4 \times 3 \times 2 \times 1$ and is therefore 4 factorial, written 4! and often read as '4 bang' even by mathematicians, or '4 shriek' by school-children.

$n!$ increases in size very rapidly indeed. 20! is already 2,432,902,008,176,640,000.

1,000,000! has recently been calculated by Harry Nelson and David Slowinski, who have several times held the record for the largest known prime. It has 5,565,709 digits and the computer printout was 5 inches high.

The factorial function turns up everywhere in mathematics. There are $n!$ different ways of arranging n objects in order.

There are $52 \times 51 \times 50 \times 49$ ways of choosing 4 cards from a pack of 52 if the order makes a difference, and choosing 4H, 3S, QD and JC for example is not the same as choosing QD, 3S, JC and 4H. $52 \times 51 \times 50 \times 49$ can be written very neatly as 52!/48!.

If the order does not make a difference, then any of the 4! = 24 ways of choosing the cards will be equivalent, and the total must be divided by 4! It can now be written 52!/48!4!.

Factorials appear in a very thin disguise in Pascal's triangle, which was used by Cardan, Tartaglia, Pascal and others to solve problems of combinations and probability as well as to calculate binomial co-efficients.

If 52!/48!4! is written, as is usual, as $\binom{52}{48}$, or as $\binom{52}{4}$ (it makes no difference!) then Pascal's triangle, on the left, can also be written as on the right: reading across the 4th line, $(x + 1)^3 = x^3 + 3x^2 + 3x + 1$.

$$
\begin{array}{ccccccc}
& & & 1 & & & & & & & \binom{0}{2} \\
& & 1 & & 1 & & & & \binom{1}{0} & & \binom{1}{1} \\
& 1 & & 2 & & 1 & & \binom{2}{0} & & \binom{2}{1} & & \binom{2}{2} \\
1 & & 3 & & 3 & & 1 & \binom{3}{0} & \binom{3}{1} & \binom{3}{2} & \binom{3}{3}
\end{array}
$$

The entries in Pascal's triangle are all integers, which illustrates the fact that the product of any n consecutive integers is always divisible by $n!$ Factorials also appear in the difference triangles for sequences of powers. Here the final differences for 5th powers are all 5!

$$
\begin{array}{cccccccc}
1 & 32 & 243 & 1024 & 3125 & 7776 & 16807 & \ldots \\
31 & 211 & 781 & 2101 & 4651 & 9031 & & \ldots \\
180 & 570 & 1320 & 2550 & 4380 & & & \ldots \\
390 & 750 & 1230 & 1830 & & & & \ldots \\
360 & 480 & 600 & & & & & \ldots \\
120 & 120 & 120 & 120 & & & & \ldots
\end{array}
$$

They also appear in the infinite series for e^x and for the trigonometrical ratios such as $\sin x$ and $\cos x$.

$$\sin x = x - x^3/3! + x^5/5! - x^7/7! + \ldots$$
$$e^x = 1 + x + x^2/2! + x^3/3! + x^4/4! + \ldots$$

Bearing in mind the connection between π, circles and the trigonometrical ratios, Stirling's formula is not unbelievable, merely astonishing: $n! \sim n^n e^{-n} \sqrt{2\pi n}$. Like Euler's relationship, it links together several important numbers and functions, though this formula is only approximate, as the $\sim$ symbol indicates.

$n!$ is just as significant in the theory of numbers. Both Wilson's theorem and its converse are true: p is prime if and only if $(p - 1)! + 1$ is divisible by p. Leibniz knew Wilson's theorem long before it was published by Edward Waring in 1770. It was Waring who ascribed it to Sir John Wilson giving him his own kind of immortality. In theory it can be used to test if a number is prime. In practice, it is an absurd test, because $n!$ is so large. It is not plausible to test if 23 is prime by dividing 23 into $22! + 1$.

The notation itself is of interest. Mathematics demands notations that are simple and striking and appropriate. What could be more appropriate than the exclamation mark for a function that increases in size so rapidly! Augustus de Morgan, the scourge of circle-squarers and other unfortunates, was most upset when the '!', which had been invented in Germany by Christian Kramp in 1808, made its way to England. He wrote

Among the worst barbarisms is that of introducing symbols which are quite new in mathematical, but perfectly understood in common, language. Writers have borrowed from the Germans the abbreviation $n!$ ... which gives their pages the appearance of expressing admiration that 2, 3, 4, etc., should be found in mathematical results. [Cajori, *A History of Mathematical Notations* v2, 1929]

If he had stopped to consider the psychology of its use, he would have appreciated that users would very quickly ignore the shock! horror! aspect and see, literally, only its mathematical meaning.

Factorials, like Fibonacci numbers, can be used as the basis for a notation for numbers that does not depend on any particular base. Simply divide the number by the largest factorial below it, then repeat with the remainder, and so on.

$2000 = (2 \times 720) + (4 \times 120) + (3 \times 24) + (1 \times 6) + (2 \times 1) = (2 \times 6!) + (4 \times 5!) + (3 \times 4!) + (1 \times 3!) + (1 \times 2!) + (0 \times 1!)$ or 243,110 in factorial.

Adding 2 such numbers is tricky and multiplication is a kind of nightmare, but they have their specialist uses. [Gruenberger, 'Computer Recreations', *Scientific American*, April 1984]

25

A square and the sum of 2 squares: $25 = 3^2 + 4^2 = 5^2$.

The Greeks represented the squares, as they did all polygonal numbers, by a pattern of dots. To turn one square into the next, it is sufficient to add a border of dots along two sides. The sizes of these borders are the odd numbers, 1, 3, 5, . . .

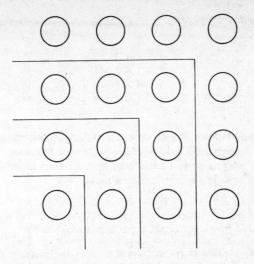

It follows that the sums of the sequence of odd numbers are the square numbers. In particular, $5^2 = 25 = 1 + 3 + 5 + 7 + 9$.

Every square is also the sum of two triangular numbers: $25 = 10 + 15$, which may be represented in a pattern of dots:

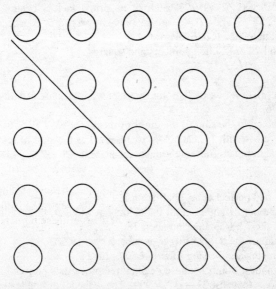

or in this number pattern:

$$
\begin{array}{ll}
1 & = 1^2 \\
1 + 2 + 1 & = 2^2 \\
1 + 2 + 3 + 2 + 1 & = 3^2 \\
1 + 2 + 3 + 4 + 3 + 2 + 1 & = 4^2 \\
1 + 2 + 3 + 4 + 5 + 4 + 3 + 2 + 1 = 5^2
\end{array}
$$

and so on . . .

Being an odd square, it is a source of the following pattern: split 25 into successive integers, $25 = 12 + 13$. Split its root likewise, $5 = 2 + 3$. Then the 3 integers up to 12 and the 2 integers from 13 have the same sums of squares. The complete pattern starts:

$$
\begin{array}{c}
3^2 + 4^2 = 5^2 \\
10^2 + 11^2 + 12^2 = 13^2 + 14^2 \\
21^2 + 22^2 + 23^2 + 24^2 = 25^2 + 26^2 + 27^2
\end{array}
$$

and so on . . . This may be compared with the pattern: $1 + 2 = 3$; $4 + 5 + 6 = 7 + 8$; $9 + 10 + 11 + 12 = 13 + 14 + 15$; and so on.

All powers of 25 end in the same digits, 25.

$25 = 4! + 1$. This is the only solution of $(n - 1)! + 1 = n^k$. [Liouville]

Fermat asserted correctly, without proving, that $25 = 3^3 - 2$ is the only square that is 2 less than a cube.

A number is *powerful* if each of its prime factors occurs at least squared. $25 = 5^2$ and $27 = 3^3$ are the only known pair of consecutive odd powerful numbers. [Golomb, AMM v77]

1, 9 and 25 are Cullen numbers and squares.

26

26 is the smallest non-palindromic number whose square is palindromic: $26^2 = 676$.

26 is equal to the sum of the digits of its cube: $26^3 = 17,576$.

26, together with 11, cannot be represented as the sum of less than 6 hexagonal numbers from the sequence 1 6 15 28 45 . . . [Guy 137]

27

The first odd perfect cube, apart from 1.

The number of points in all the colours at snooker, because it is one less than the 7th triangular number, 28.

The sum of the digits of its own cube: $27^3 = 19,683$.

All integers are the sum of at most 27 primes.

The smallest number whose decimal reciprocal has period 3: $1/27 = 0 \cdot 037037 \ldots$ (because $27 \times 37 = 999$). The sequence of smallest numbers

with reciprocals of period n starts: 3 11 27 101 41 7 239 73 81 451 ...
[Sloane 2886]

If a 3-digit multiple of 27 is permuted cyclically, so that for example 513 turns into 135 or 351, then the resulting number is still a multiple of 27. The only other number with this property for 3-digit numbers is 37.

27 is the smallest number that is 3 times the sum of its digits.

27 is the smallest integer that is the sum of 3 squares in 2 ways: $27 = 3^2 + 3^2 + 3^2 = 5^2 + 1^2 + 1^2$.

The '$3x + 1$' problem starts with any number, divides it by 2 if it is even, and multiplies it by 3 and adds 1 if it is odd. This process is then repeated. For example, the sequence starting with 17 runs: 17 52 26 13 40 20 10 5 16 8 4 2 1.

All the integers less than 1,000,000,000 have been tested, and every one eventually ends in the sequence $4-2-1$. It is not known whether every number eventually reaches 1. Of the first 50 integers, 27 takes the longest, 111 steps, reaching a maximum height of 9232.

The numbers which set records for the number of halving and tripling steps to reach 1, are:

number	1	2	3	6	7	9	18	25	27
steps	0	1	7	8	16	19	20	23	111

The first few numbers to take a large number of tripling steps to reach 1, all take either 38, 39, 40 or 41. 27 is the first and takes 41.
[Sloane 4323]

There are 27 ways to dissect a nonagon into triangles, with 6 chords, where dissections which are equivalent by rotation or reflection are not distinguished. Starting with a triangle, the number of ways are: 1, 1, 1, 3, 4, 12, 27, 82, 228, 733, 2282 ... (If rotations and reflections are distinguished, the number of ways is given by the Catalan sequence.)
[Sloane 2375]

28

The number of days in the lunar cycle.

In the imperial system of weights and measures, the number of pounds in a quarter.

The 7th triangular number and the number of dominoes in a standard double-six set.

The longest known sociable chain is of 28 links, starting with 12,496.

Perfect numbers

28 is the second perfect number, following 6, meaning that 28 is the sum

of its divisors, including unity but excluding itself: $28 = 1 + 2 + 4 + 7 + 14$.

The first 4 perfect numbers, 6, 28, 496 and 8218, were known to the late Greeks. Nicomachus and Iamblichus listed all 4 and Iamblichus, not unnaturally bearing in mind that he had no conception of the number base 10 as mathematically arbitrary, conjectured that there was one perfect number for each number of digits, and further that they not only ended in either 6 or 8, which is true, but that the 6s and 8s alternate, which is not. The sequence of unit digits actually goes $6-8-6-8-6-6-8-8-6-6-8-8-6-8-8$ ... and every perfect number ends in either 28 or 6 preceded by an odd digit.

Euclid, whose *Elements* is not limited to geometry, proved in Book IX that, 'If as many numbers as we please beginning from a unit be set out continuing in double proportion until the sum of all becomes prime, and if the sum multiplied into the last make some number, the product will be perfect.'

In other words, if, for example, $1 + 2 + 4 + 8 + 16$ is prime, which it is, being $2^5 - 1 = 31$, then 31×16 is perfect. In fact it is 496, the third perfect number. $28 = 2^2(2^3 - 1)$ and $6 = 2(2^2 - 1)$.

In each case the bracketed factors, $2^3 - 1$ and $2^2 - 1$, which are Mersenne numbers, are prime. This is the critical condition.

Euclid proved only that his rule was sufficient. It was Euler, 2000 years later, who proved that all even perfect numbers (odd perfect numbers are quite a different matter) are of the form $2^{n-1}(2^n - 1)$ where $2^n - 1$ is a Mersenne prime M_n.

Every even perfect number is hexagonal and also therefore triangular. Only 28 is the sum of two equal powers: $28 = 3^3 + 1^3$.

It follows from the definition of perfection that the sum of reciprocals of the divisors of a perfect number is 2. For example, since $28 + 28 = 1 + 2 + 4 + 7 + 14 + 28$, we can divide through by 28 and obtain: $2 = 1/28 + 1/14 + 1/7 + 1/4 + 1/2 + 1/1$.

Less obviously, every even perfect number, except 6, is a partial sum of the series $1^3 + 3^3 + 5^3 + 7^3 + 9^3 + \ldots$ For example, $28 = 1^3 + 3^3$ while $496 = 1^3 + 3^3 + 5^3 + 7^3$.

With the same exception, 6, it also follows from Euclid's formula that the digital root of an even perfect number is 1, or, which amounts to the same thing, that every perfect number leaves the remainder 1 when divided by 9.

The perfect numbers correspond one-for-one with the Mersenne primes, whose history is sketched under *127*. As long as only hand calculation was available, the discovery of Mersenne primes depended on human labour in actually making the necessary calculations, and subtle theorems that showed that only possible divisors of a certain type need be tried. The labour for large numbers was immense. Mersenne himself stated that all eternity would not be sufficient to decide if a 15- or 20-digit number were prime. In 1814 Peter Barlow in an article in *A New Mathematical and Philosophical Dictionary* wrote,

Euler ascertained that $2^{31} - 1 = 2,147,483,647$ is a prime number; and this is the greatest at present known to be such, and, consequently, the last of the above perfect numbers, which depends upon this, is the greatest perfect number known at present, and probably the greatest that ever will be discovered; for, as they are merely curious without being useful, it is not likely that any person will attempt to find one beyond it. [Shanks, *Solved and Unsolved Problems in Number Theory* v1, 1962]

Barlow underestimated the fascination of record-breaking for mathematicians, and he could not foresee the electronic computer. By allowing millions of calculations per second, the computer opened up vast reaches of numbers that had previously been inaccessible and allowed mathematicians to make effective use of much more powerful tests for primality. These tests decide whether n is prime, by analysing the factors of either $n - 1$ or $n + 1$.

Because of their special form, Mersenne and Fermat numbers are easier to test for primality than any other forms, and all the recent record-breaking primes have been Mersenne numbers, and have automatically led to a new perfect number.

Overleaf is a complete list of the perfect numbers known to date: M_p stands for the Mersenne prime $2^p - 1$. **Bold** type indicates that either the perfect number itself, or its associated Mersenne prime, is an entry in the main body of this dictionary. The question marks after entries 32 and 33 indicate that it is not yet certain that no smaller perfect numbers remain to be found.

Is the number of perfect numbers infinite? The table shows perfect numbers occurring less and less frequently, with some surprising jumps. The largest is from the 12th to the 13th, where the index of the Mersenne prime jumps from 127 to 520, a more than fourfold increase in index. The index nearly doubles again from the 14th to the 15th to the 16th: 607 to 1279 to 2203, from the 23rd to the 24th, and from the 28th to the 29th. This suggests that perfect numbers thin out pretty quickly, but it says nothing about their total number. They could disappear completely – or there could be many more of them up among the unimaginably large numbers. (And most integers are indeed unimaginably large.)

Odd perfect numbers are a curiosity in themselves. Guy describes the existence of odd perfect numbers as one of the more notorious unsolved problems in number theory.

1	$2M_2$	6	known to Greeks
2	2^2M_3	28	known to Greeks
3	2^4M_5	496	known to Greeks
4	2^6M_7	8128	known to Greeks
5	$2^{12}M_{13}$	33550336	recorded in medieval manuscript
6	$2^{16}M_{17}$	8589869056	Cataldi, 1588, $M_{17} = 131,071$
7	$2^{18}M_{19}$	137438691328	Cataldi, 1588, $M_{19} = 524,287$
8	$2^{30}M_{31}$		Euler, 1750, $M_{31} = 2,147,483,647$
9	$2^{60}M_{61}$		Pervushin, 1883
10	$2^{88}M_{89}$		Powers, 1911
11	$2^{106}M_{107}$		Powers and Fauquembergue, 1914
12	$2^{126}M_{127}$		Lucas, 1876
13	$2^{520}M_{521}$		Robinson, 1952
14	$2^{606}M_{607}$		Robinson, 1952
15	$2^{1278}M_{1279}$		Robinson, 1952
16	$2^{2202}M_{2203}$		Robinson, 1952
17	$2^{2280}M_{2281}$		Robinson, 1952
18	$2^{3216}M_{3217}$		Riesel, 1957
19	$2^{4252}M_{4253}$		Hurwitz, 1961
20	$2^{4422}M_{4423}$		Hurwitz, 1961
21	$2^{9688}M_{9689}$		Gillies, 1963
22	$2^{9940}M_{9941}$		Gillies, 1963
23	$2^{11212}M_{11213}$		Gillies, 1963
24	$2^{19936}M_{19937}$		Tuckerman, 1971
25	$2^{21700}M_{21701}$		Nickel and Noll, 1978
26	$2^{23208}M_{23209}$		Noll, 1979
27	$2^{44496}M_{44497}$		Nelson and Slowinski, 1979
28	$2^{86242}M_{86243}$		Slowinski, 1982
29	$2^{110502}M_{110503}$		Colquitt and Welsch, 1988
30	$2^{132048}M_{132049}$		Slowinski, 1983
31	$2^{216090}M_{216091}$		Slowinski, 1985
32?	$2^{756838}M_{756839}$		Slowinski and Gage, 1992
33?	$2^{859432}M_{859433}$		Slowinski and Gage, 1994

However, researchers, without having produced any odd perfects, have discovered a great deal about them, if it makes sense to say that you know a great deal about something that may not exist.

Descartes claimed that an odd perfect is a product of a square and a prime. Euler proved that an odd perfect must be of the form $p^a q^b r^c \ldots$ where the p, q, $r \ldots$ are all of the form $4n + 1$, a is of the same form, and b, $c \ldots$ are all even.

Coming to modern times, there are no odd perfect numbers less than 10^{300}. An odd perfect must have at least 8 distinct prime factors (11 if it is not divisible by 3), and must be divisible by a prime power greater than 10^{18}. The greatest prime factor must be greater than 300,000 and the second largest must be greater than 1000. Any odd perfect less than 10^{9118} is divisible by the 6th power of some prime.

29

No sum of three 4th powers is divisible by either 5 or 29 unless they all are. [Euler]

29 is the third number n, following 1 and 5, such that $2n^2 - 1$ is a square: $2 \times 29^2 - 1 = 41^2$.

The quadratic form $2x^2 + p$, with $p = 3, 5, 11$ or 29, gives prime values for $x = 0, 1, \ldots p - 1$. [Ribenboim 142]

Not more than 3 successive numbers can be square-free, because every 4th number is divisible by 4. However, 29 to 43, omitting 32, 36 and 40, is the first sequence of 4 successive triplets, all of them square-free. The next sequence is from 101 to 115. [Sloane 3824]

$29 = (2 \times 3 \times 5) - 1 = \text{primorial } (5) - 1$.

Primorial $(n) - 1$ is prime for 3, 5, 11, 13, 41, 89, 317, 991, 1873, 2053, and no other values below 2377. [Buhler, Crandall and Penk, MOC v38]

30

Primorials

Primorial p, denoted by $p\#$, is defined only if p is prime. It is then equal to the product of all the primes up to and including p. $30 = 5\# = 5 \times 3 \times 2$.

The sequence of primorials starts: 1 2 6 30 210 2310 30,030 510,510 9,699,690 223,092,870 . . .

30 is the smallest integer with 3 distinct prime divisors, 2, 3 and 5. The smallest with 4 distinct prime divisors is $7\# = 2 \times 3 \times 5 \times 7 = 210$, and so on.

Most integers have very few distinct prime divisors. The average for numbers less than 100 is about 1·71. For numbers less than 100,000,000 the average is only about 2·9. For 10^{100}, a googol, the average is still only about 5·4, and for 10^{googol} it has risen only to about 231. Primorials are exceptional in this respect.

30 is the greatest number such that all the numbers less than it and prime to it are themselves primes. The other numbers with this property are 2, 3, 4, 6, 8, 12, 18 and 24.

There are only 2 Pythagorean triangles whose areas equal their perimeter. One is the $5-12-13$ triangle whose area and perimeter are both 30. The other is the $6-8-10$ triangle, whose area and perimeter are both 24.

30 is the area of the smallest rectangle on which a re-entrant knight's

tour is possible. It can be done on either a 5-by-6 or on a 3-by-10 board. The smallest square board is 6 by 6.

The dodecahedron and its dual, the icosahedron, each have 30 edges.

30 is the smallest number which has not been represented as the sum of 3 integer cubes. [Guy 151]

$1/2 + 1/3 + 1/5 - 1/30 = 1$, so 30 is a Guiga number, the smallest. The next is $858 = 2 \times 3 \times 11 \times 13$, because $1/2 + 1/3 + 1/11 + 1/13 - 1/858 = 1$. [AMM v103 45]

The arithmetic progression $30-66-102-138$ is the smallest of four terms consisting of numbers each with three distinct primes.

$30\frac{1}{4}$

There are $30\frac{1}{4}$ (= $5\frac{1}{2}$ squared) square yards in a square rod, pole or perch, or $272\frac{1}{4}$ square feet.

31

$2^5 - 1$. The 5th Mersenne number and the 3rd Mersenne prime, leading to the 3rd perfect number, 496.

$$31 = 1 + 5 + 5^2 = 1 + 2 + 2^2 + 2^3 + 2^4$$

One of only 2 known numbers that can be written in 2 ways as the sum of successive powers, starting from 1. The other is 8191.

The first prime number the decimal period of whose reciprocal is an odd number of digits in length.

$1/31 = 0 \cdot 03225\ 80645\ 16129\ 03225 \ldots$

Note these products:

$032258 \times 2 =$	$64,516$	$032258 \times\ 9 = 290,322$
$032258 \times 4 =$	$129,032$	$032258 \times 14 = 451,612$
$032258 \times 5 =$	$161,290$	$032258 \times 16 = 516,128$
$032258 \times 7 =$	$225,806$	$032258 \times 18 = 580,644$
$032258 \times 8 =$	$258,064$	$032258 \times 19 = 612,902$

and so on . . .

Note also that $03225 + 80645 + 16129 = 99,999$ and $032 + 258 + 065 + 416 + 129 = 900$.

$(2 \times 3 \times 5) + 1 = 5\# + 1$

$n\# + 1$ is prime for 2, 3, 5, 7, 11, 31, 379, 1019, 1021, 2657, and for no other values below 3088.

Start with any number, and square and add its digits. Then repeat. If you eventually get to 1, you started with a happy number. 31, 32 is the first pair of consecutive happy numbers.

If n points are marked irregularly on the circumference of a circle, and

[92]

all the diagonals drawn, so that no three diagonals concur, then the number of regions created is 1, 2, 4, 8, 16, 31 . . .

Two binary puzzles
The Tower of Hanoi was brought out by Edouard Lucas under the name M. Claus in 1883, and provided a year later with a charming but wholly fictitious story:

In the great temple at Benares, beneath the dome which marks the centre of the world, rests a brass plate in which are fixed three diamond needles, each a cubit high and as thick as the body of a bee. On one of these needles, at the creation, God placed 64 discs of pure gold, the largest disc resting on the brass plate, and the others getting smaller and smaller up to the top one. This is the Tower of Bramah. Day and night unceasingly the priests transfer the discs from one diamond needle to another according to the fixed and immutable laws of Bramah, which require that the priest on duty must not move more than one disc at a time and that he must place this disc on a needle so that there is no smaller disc below it. When the 64 discs have been thus transferred from the needle on which at the creation God placed them to one of the other needles, tower, temple, and Brahmins alike will crumble into dust, and with a thunderclap the world will vanish.

This account of Hindu theology is nonsensical, but the problem itself has a very neat solution involving powers of 2. In the figure the 5 rings on one peg are to be transferred to one of the other pegs.

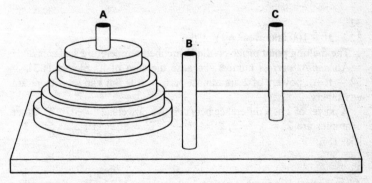

It will be found by trial and error that to transfer 1, 2 or 3 rings requires respectively 1, 3 and 7 moves. In general, to move $n + 1$ rings to peg A from peg B, requires that n rings be moved to peg C, that the largest ring be then removed to B, and the first n rings then moved to B by a repetition of the previous sequence. So the number of moves required is each time one more than double the previous total.

The sequence therefore continues 1 3 7 15 31 63 and the general

[93]

term is $2^n - 1$. The rings in the figure require 31 moves. In practice it helps mentally (or physically!) to mark the rings as alternately odd and even; if there are an odd number of rings to be moved, the first move should be on to the target peg, if even, on to the third peg.

The Brahmin priests would need $2^{64} - 1$ or 18,446,744,073,709,551,615 moves to complete their task, or nearly 600,000,000,000 years at one move per second all day and every day.

Fibonacci considered the problem of the smallest set of weights required to weigh any weight up to a given amount. Tartaglia solved the problem when only one pan may be used, and Bachet solved it when one or both pans may be employed. If the weights are placed in one pan of the balance only, the maximum weight that can be weighed with only 5 weights is 31, by using the values 1, 2, 4, 8 and 16. In general, n weights from this sequence will weigh up to $2^n - 1$.

Each weighing is performed in a unique manner represented by the value expressed in binary. To weigh 26, which is 11010 in binary, we use the 16, the 8 (not the 4) and the 2 (but not the unit) weights.

Using both pans, the solution is similar, but now relies on expressing the weight as the sum and difference of powers of 3. With the weights 1, 3, 9 and 27 it is possible to weigh up to 40. In general the weights up to 3 will weigh up to a maximum of $\frac{1}{2}(3^{n+1} - 1)$.

32

$32 = 2^5 = 100,000$ in binary notation.

The melting point of ice on the Fahrenheit temperature scale.

An *almost perfect* number, because the sum of its factors is $31 = 32 - 1$. All powers of 2 are almost perfect. It is not known if there are any others.

A power of 2 as a difference between two powers: $2^5 = 3^4 - 7^2$. Other examples are $2^0 = 3^2 - 2^3$; $2^1 = 3^3 - 5^2$; $2^2 = 5^3 - 11^2$; $2^4 = 5^2 - 3^2$. [Guy 156]

33

A semi-prime is a number with only 2 factors. 33–34–35 is the smallest triplet of successive semi-primes.

$33 = 1! + 2! + 3! + 4!$

33 is the largest number that is not the sum of distinct triangular numbers.

Neither 2^{33} nor 5^{33} in base 10 contains any zero. 33 is probably the largest such number. [Sloane 497]

A palindrome in base 10 and also in base 2. The sequence of such

numbers starts: 1   3   5   7   9   33   99   313   585   717   7447   9009
15,351   32,223   39,993 ... [JRM v18 47]

Any integer greater than 33 can be written as the sum of 5 non-zero
squares. [Jackson, Masat, Mitchell, MM v61 41]

The triplet 33, 34, 35 is the smallest in which each number is the
product of 2 distinct primes. Therefore they each have 4 divisors, and are
multiplicatively perfect: that is, the product of the divisors of n is n^2. The
other numbers with these three properties are the cubes of primes. The
next such triplets are: 93, 94, 95; 141, 142, 143; 201, 202, 203; 213, 214,
215; 217, 218, 219; ...

The smallest number such that $\sigma(n) = \sigma(n + 2)$. (Both equal 48.)

Lucky numbers
The prime numbers can be found by using the Sieve of Eratosthenes:
write down the integers in order and strike out every other number, to
get rid of the multiples of 2. Then strike out every 3rd number in the
original sequence to get rid of multiples of 3, and so on. Lucky numbers
are constructed by a similar process. First strike out every other number,
leaving the odd numbers:
    1   3   5   7   9   11   13   15   17   19 ...
After 1, the next odd number is 3, so strike out every 3rd number in
this sequence, leaving:
    1   3   7   9   13   15   19 ...
The next number remaining is 7, so strike out every 7th number, starting
with 19. And so on.

The Lucky numbers are those that remain. The sequence starts:
    1   3   7   9   13   15   21   25   31   33   37   43   49   51 ...

Lucky numbers share many properties with the prime numbers, which
suggests that those properties, surprisingly, belong to the primes not
because each prime has no factors but itself and 1, but because of the
way in which the primes can be constructed by the Sieve of Eratosthenes.
It is likely that any sequence constructed by a similar sieve will have the
same properties. [Guy]

34
The magic constant of a 4 × 4 magic square.

The smallest number to be the sum of 2 Lucky numbers in 4 ways.

35
There are 35 hexominoes, each formed of 6 squares attached edge to
edge. Surprisingly, although the total area of the 35 hexominoes is 210,
which might make a rectangle 3 × 70 or 5 × 42 or 6 × 35 or 7 × 30 or

10×21 or 14×15, not one of these rectangles can actually be filled with the 35 pieces.

After hexominoes, the number of n-ominoes rises rapidly. There are 108 heptominoes, of which 1 has a hole. There are 369 octominoes, 6 having a hole, and 1285 9-ominoes, of which 37 have a hole.

The maximum length of a knight's tour on a standard chessboard which does not cross itself is 35 moves.

35 and 4375 have the same prime factors between them (namely, 2, 3, 5 and 7) as have 36 and 4374. [Guy 75]

Pascal's triangle
The numbers in Pascal's triangle are so important that they had to be included, although no one of them is more typical of the triangle than any other, so I have chosen 35 as their representative.

```
                    1
                  1   1
                1   2   1
              1   3   3   1
            1   4   6   4   1
          1   5   10   10   5   1
        1   6   15   20   15   6   1
      1   7   21   35   35   21   7   1
    1   8   28   56   70   56   28   8   1
  1   9   36   84   126   126   84   36   9   1
```
. . . .

The triangle is named after Blaise Pascal, the brilliantly precocious mathematician, natural scientist and theologian who wrote a *Treatise on the Arithmetical Triangle*. Pascal, however, was the last rather than the first of many mathematicians who considered almost identical arrays in connection with the extraction of roots, problems in probability and combinations, and the calculation of binomial coefficients. He brought together and built on their results.

Chu Shih-chieh in *The Precious Mirror of the Four Elements* (1303) gives a pyramidal array identical to our modern arrangement, for determining one binomial coefficient from another. [David, *Games, Gods and Gambling*, OUP, 1987]

It was first published in Europe in 1529 and was given in varying forms by, among others, Stifel, Tartaglia, who used it to calculate the expansion of a 12th power, and Cardan, who used it in problems of combinations and polygonal numbers, and by Herigone, who is supposed to have been Pascal's teacher, and whose works Pascal himself cites.

Pascal defined the triangle by stating that each cell is occupied by the sum of cells above it, the edge cells being unity. He first deduced 19 consequences, or, as we should say, theorems, including, 'In every triangle the sum of the cells of each base is a number of the double progression beginning with unity . . .' In other words, the sum of the numbers in row n is 2^n.

(The top row, which is sometimes omitted, counts as row 0.)

He went on to study 'orders of numbers' and the use of the triangle in calculating combinations, the division of stakes between gamblers, and binomial coefficients. The 'orders of numbers' were the diagonal sequences. The first diagonal is occupied by units, the second by the natural numbers. The 3rd diagonal is the triangular numbers, 1, 3, 6, 10 . . . and the 4th is the tetrahedral numbers, 1, 4, 10, 20, 35 . . . which are the numbers of cannonballs needed to stack into triangular pyramids of increasing size. The subsequent diagonals can be interpreted as arrangements of higher dimension, from the 4th upwards, though Pascal possessed no such modern conception.

To calculate 'combinations', for example the number of ways of selecting 3 dishes from a menu of 7, it is necessary only to go to the 4th number in the 7th line: it is 35, or in modern notation $\binom{7}{3} = 35$ which can be calculated as $(7 \times 6 \times 5)/(3 \times 2 \times 1)$ or as $7!/3!4!$.

Binomial coefficients are found in the same way. The coefficient of x^3 in the expansion of $(1 + x)^7$ is $\binom{7}{3}$ or 35, and the complete expansion is:

$$(1 + x)^7 = 1 + 7x + 21x^2 + 35x^3 + 35x^4 + 21x^5 + 7x^6 + x^7$$

From the rule for constructing the triangle it follows that

$$\binom{n}{m} \quad \binom{n}{m+1} \quad \binom{n+1}{m+1},$$

which Pascal expressed in the language of orders.

The triangle has many other features. Pascal himself wrote, '. . . I leave out many more than I include; it is extraordinary how fertile in properties this is. Everyone can try his hand.'

The entries in row p, except the units, are divisible by p if and only if p is prime.

The number of odd entries in any row of Pascal's triangle is always a power of 2.

The shallow diagonals, 1, 1–1, 1–2, 1–3–1, 1–4–3, 1–5–6–1, 1–6–10–4 . . . sum to the Fibonacci sequence, 1 1 2 3 5 8 . . .

There are an infinite number of rows containing 3 numbers in arithmetical progression, such as 7–21–35. The next two such sets are 1001–2002–3003, and 490,314–817,190–1,144,066. On the other hand, there are no triplets of numbers forming geometric or harmonic progressions. [Motzkin, *Scripta Mathematica* v12]

There is, however, a neat connection with the harmonic series, the reciprocals of the natural numbers:

$$1 = 1$$
$$1 - 1/2 = 1/2$$
$$1 - (2 \times 1/2) + 1/3 = 1/3$$
$$1 - (3 \times 1/2) + (3 \times 1/3) - 1/4 = 1/4$$
$$1 - (4 \times 1/2) + (6 \times 1/3) - (4 \times 1/4) + 1/5 = 1/5$$

and so on.

There are further connections with the Harmonic Triangle of Leibniz:

$$
\begin{array}{ccccccccccc}
 & & & & & \frac{1}{1} & & & & & \\
 & & & & \frac{1}{2} & & \frac{1}{2} & & & & \\
 & & & \frac{1}{3} & & \frac{1}{6} & & \frac{1}{3} & & & \\
 & & \frac{1}{4} & & \frac{1}{12} & & \frac{1}{12} & & \frac{1}{4} & & \\
 & \frac{1}{5} & & \frac{1}{20} & & \frac{1}{30} & & \frac{1}{20} & & \frac{1}{5} & \\
\frac{1}{6} & & \frac{1}{30} & & \frac{1}{60} & & \frac{1}{60} & & \frac{1}{30} & & \frac{1}{6}
\end{array}
$$

and so on.

Each fraction here is the sum of the numbers immediately below it. The terms in each row are the initial term divided by the corresponding Pascal triangle entries.

Each entry is the sum of the infinite series that starts immediately below it to the left and continues downwards along the diagonal to the right, for example: $1/4 = 1/5 + 1/30 + 1/105 + \ldots$

In Pascal's triangle, each number is the sum of either of the diagonals starting immediately above it, and taking the long way to the edge: for example, $35 = 15 + 10 + 6 + 3 + 1$. (So the sum of the first 5 triangular numbers is 35.) The first few rows of Pascal's triangle may give the impression that almost all its entries are different, apart from the left–right symmetry and the edge units. This is not so, as the appearance of three 6s and four 10s might suggest. (*See 3003.*)

36

The 8th triangular number, thought of by the Greeks as also the sum of the first 4 even numbers and first 4 odd numbers.

It is also square, and the first number after 1 to be both square and triangular. The numbers that are both square and triangular are beautifully related to the best approximations to $\sqrt{2}$:

number	root	factors of the root
1	1	1×1
36	6	2×3
1225	35	5×7
41616	204	12×17

and so on.

In each case the factors of the root are the numerator and denominator of the next approximation to $\sqrt{2}$.

Because its square root is the 3rd triangular number, it is also the sum of the first 3 cubes: $36 = 1^3 + 2^3 + 3^3$.

$1 - 6 - 36$ is the first set of triangular numbers in geometrical progression.

The sequence of squares which are values of $\sigma(n)$ goes: 1 4 36 121 144 256 576 . . .

36 is the largest 2-digit number divisible by the product of its digits and by their sum.

Every sequence of 7 consecutive numbers greater than 36 includes a multiple of a prime greater than 41. [Gupta, *Selected Topics in Number Theory*, Abacus Press, 1980]

37

Any 3-digit multiple of 37 remains a multiple when its digits are cyclically permuted.

Every number is the sum of at most 37 5th powers.

The juggler sequence is defined by $n \rightarrow [n^{1/2}]$ if n is even, and $n \rightarrow [n^{3/2}]$ if n is odd. The problem is, do all juggler sequences finally end in 1? 37 is the first tall peak in its graph: the sequence starting with 37 rises to 24,906,114,455,136 before falling back to 1 after 18 steps. [Pickover, *Computers and the Imagination*, 1991]

The reciprocal, $1/37 = 0.027\ 027\ .\ .\ .$ has period only 3.

The 4th centred hexagonal number, obtained by arranging hexagonal layers of points around a central point.

The formula for the *n*th centred hexagonal number is $3n(n-1) + 1$. By a different division of the original diagram the *n*th centred hexagonal number is equal to $6T_{n-1} + 1$, where T_n is the *n*th triangular number.

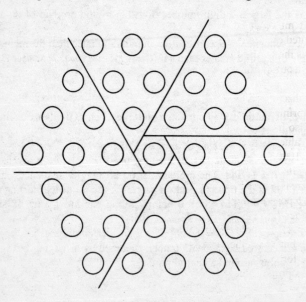

38

The magic constant in the only possible magic hexagon, which uses the numbers 1 to 19.

The largest even integer that cannot be expressed as the sum of 2 composite odd numbers. More generally, the largest even integer that cannot be written as the sum of $2k$ composite odd integers, is $18k + 20$. [Ruemmler and Shippensburg, MM v63 276]

$n + d(n) = 38$ for $n = 30, 32$ and 34, the smallest such triple.

39

Like all 2-digit numbers ending in 9, it equals the product of its digits plus their sum.

The number of convex polygons that can be assembled from a complete set of 12 hexiamonds. [Golomb, *Polyominoes*, 1966]

40

Equal to two score. A biblical expression for a long period of time, for example, 40 days in the wilderness, 40 years of wandering in the desert.

There are 40 rods, perches or poles in a furlong of 220 yards.

The name 'forty' is the unique number name in which the letters appear in alphabetical order. [Richard Harris]

41

5-digit multiples of 41 remain multiples of 41 when their digits are permuted cyclically.

41 is the smallest prime which is not the difference, in some order, between a power of 2 and a power of 3. [Gauchman and Rosenholtz, MM v66 269]

Euler discovered the excellent and famous formula $x^2 + x + 41$, which gives prime values for $x = 0$ to 39.

It also has many further prime values, including 581, among its first 1000 values. However, other quadratic formulae can beat this: for example, $2x^2 - 1000x - 2609$ has 602 prime values (including negative values) among its first 1000. [Guy 37]

There is no quadratic formula of the form $x^2 + ax + b$, with coefficients a and b positive and less than 10,000, which produces a longer sequence of primes.

The '41' formula also gives prime values for negative values from -1 to -40 but this is the same set of primes repeated.

The formula $x^2 - 79x + 1601$ is just a variation on the '41' formula. It gives prime values for $x = 0$ to 79, repeating each prime once.

If $q = 2, 3, 5, 11, 17$ or 41, then $x^2 + x + q$ has prime values for $x =$

0, 1, ... $q - 2$. These are the only values of q for which this is true.
[Riebenboim 137]

When Charles Babbage built a small trial version of his Analytical Engine he calculated a list of values of Euler's function to show off its powers. At one demonstration, it is recorded,

THIRTY-TWO numbers of the same table were calculated in the space of TWO MINUTES AND THIRTY SECONDS; and as these contained EIGHTY-TWO figures, the engine produced thirty-three figures every minute, or more than one figure in every two seconds. On another occasion it produced FORTY-FOUR figures per minute. This rate of computation could be maintained for any length of time; and it is probable that few writers are able to copy with equal speed for many hours together. [Brewster, *Letters on Natural Magic*, 1856]

The writers referred to are the copyists who noted the figures that the machine produced. Of course, long after Babbage's brilliant experiments, and long after the development of the desk calculator, calculating prodigies were far faster than any machine. Indeed, if the time taken to instruct the machine is, very reasonably, taken into account, they could beat early computers on many problems.

42

A solution of $\phi(n)d(n) = \sigma(n)$. Three smaller solutions are $n = 1, 3$ or 14.

Catalan numbers

42 is the 5th Catalan number.

The sentence starts: 1 2 5 14 42 132 429 1430 4862 16,796 58,786 208,012 742,900 2,674,440 ...

The formula for the nth term is $\dfrac{1}{n + 1}\dbinom{2n}{n}$. (The sequence sometimes starts with an extra 1, thus: 1 1 2 5 14 ... in which case the formula must be adjusted accordingly.)

Literally hundreds of sequences have been studied in the solution of mathematical problems, or have been studied for their own sake.

Catalan numbers are probably the most frequently occurring combinatorial numbers, after the binomial coefficients.

In how many ways can a regular n-gon be divided into $n - 2$ triangles, if different orientations are counted separately? The answer is the Catalan sequence.

In how many ways can brackets be placed round a sequence of $n + 1$ letters, so that there are two letters inside each pair of brackets?

ab in 1 way: (*ab*)

abc in 2 ways: (*ab*)*c*   *a*(*bc*)

abcd in 5 ways: (*ab*)(*cd*)   *a*((*bc*)*d*)   ((*ab*)*c*)*d*   *a*(*b*(*cd*))   (*a*(*bc*))*d*

and so on.

In how many ways can you move on a graph from the origin to $(2n + 2, 0)$ with diagonal steps, never touching the *x*-axis, except at the start and finish? Catalan once more.

In how many ways can *n* votes be cast between two candidates, so that one chosen candidate is never behind in the counting? The answer to every one of these problems is the sequence of Catalan numbers, demonstrating that these apparently very different problems are, in a very useful sense, equivalent to each other.

43

The sequence defined by $x_n = (1 + x_0^2 + x_1^2 + \ldots + x_{n-1}^2)/n$, $n = 1, 2 \ldots$ with $x_0 = 1$ produces integers up to x_{43}, which is not an integer. The sequences starts: 2 3 5 10 28 154 3520 1,551,880 267,593,772,160 . . .

The analogous sequence with cubes instead of squares is integral up to and including x_{88}. [Guy 214]

In base 2, 43 = 101011. This base 2 number never becomes a palindrome by the reverse-and-add process. [MG v78 313–4]

$1/2 + 1/3 + 1/7 + 1/43 = 1 - 1/1806$. This is the best approximation from below to 1 as a sum of a minimal number of unit fractions. In the sequence (1) 2 3 7 43 . . . each number is the product of all the previous terms, plus 1.

44

Euler's solution to the problem of finding a brick with integral edges and face diagonals is 44, 117 and 240.

The lengths of the diagonals of the faces are 267, 125 and 244. The length of the space diagonal is not an integer. The problem of finding an integral brick in which all the diagonals are integral remains unsolved.

The largest number *n* such that the numbers 1 to *n* can be split into 4 sets so that no number in any set is the sum of two other numbers in the same set: {1, 3, 5, 15, 17, 19, 26, 28, 40, 42, 44} {2, 7, 8, 18, 21, 24, 27, 33, 37, 38, 43} {4, 6, 13, 20, 22, 23, 25, 30, 32, 39, 41} {9, 10, 11, 12, 14, 16, 29, 31, 34, 35, 36}

44 45 46 47 is the first sequence of 4 consecutive integers with falling values of σ(*n*), 84 78 72 48, respectively. The next is: 104 105 106 107, with σ(*n*) values 210 192 162 108, respectively.

With 45, the first pair of consecutive numbers each with 6 divisors. The sequence of such pairs continues: 75 76 98 99 116 117 147 148 171 172 242 243 244 245 332 333 . . .

Subfactorials
Subfactorial $5 = 5!(1 - 1/1! + 1/2! - 1/3! - 1/4! - 1/5!) = 44$

Nikolaus Bernoulli first considered the problem that may be expressed like this: n letters are written to different addresses, and n matching envelopes prepared. In how many ways can the letters be placed in the envelopes so that every letter is in the wrong envelope?

The answer is subfactorial n.

The sequence starts: 0 1 2 9 44 265 1854 14,833 . . .

45

The 3rd smallest Kaprekar number, after 1 and 9.

Also a Kaprekar 'triple', because $45^3 = 91,125$ and $9 + 11 + 25 = 45$. The sequence of such triples starts: 1 8 10 45 297 2322 . . .

Every number greater than 45 is the sum of distinct primes greater than 11. [Gupta, *Selected Topics in Number Theory*, Abacus Press, 1980]

Polygonal numbers
45 is the 5th hexagonal number, which can be calculated from the formula $n(2n - 1)$ when $n = 5$.

The sequence of hexagonal numbers starts: 1 6 15 28 45 . . .

45 is also the smallest hexamorphic number, apart from 1, because it is the 5th hexagonal number and it ends in 5. The sequence of hexamorphic numbers includes 1, and continues (45), $H(6) = 66$, $H(25) = 1225$, $H(26) = 1326$, $H(50) = 4950$ . . . [Pickover, *Computers and the Imagination*, 1991]

Polygonal numbers were studied by the Greeks. They were a natural development from triangular and square numbers, and can also be represented by patterns of dots (*opposite*).

Polygonal numbers can be constructed by drawing similar patterns, but with a larger number of sides. There are formulae for each polygonal sequence, and these also form a pattern:

name	formula	$n =$	1	2	3	4	5	6	7	. . .
triangular	$\frac{1}{2}n(n + 1)$		1	3	6	10	15	21	28	. . .
square	$\frac{1}{2}n(2n - 0)$		1	4	9	16	25	36	49	. . .
pentagonal	$\frac{1}{2}n(3n - 1)$		1	5	12	22	35	51	70	. . .
hexagonal	$\frac{1}{2}n(4n - 2)$		1	6	15	28	45	66	91	. . .
heptagonal	$\frac{1}{2}n(5n - 3)$		1	7	18	34	55	81	112	. . .
octagonal	$\frac{1}{2}n(6n - 4)$		1	8	21	40	65	96	133	. . .
and so on.										

Typically, the number 1 is simultaneously triangular, square, pentagonal . . . and so on for ever!

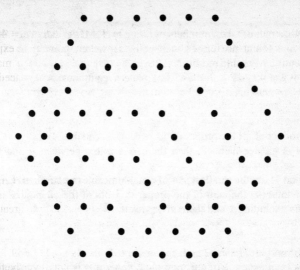

The hexagonal numbers are equal to the alternate triangular numbers. All even perfect numbers are hexagonal and therefore triangular also.

There are other obvious patterns in this table. The vertical differences are constant in each column. Thus, the *5th* numbers of each order differ by 10 which is the *4th* triangular number, and they are all divisible by 5: 15, 25, 35, 45, 55, 65, 75 . . .

The horizontal differences are the natural numbers for the triangular numbers; the odd numbers for the squares; the sequence 4, 7, 10, 13 . . . for the pentagonal numbers; 5, 9, 13, 17 . . . for the hexagonal; and so on.

Take any square of entries, say 18 – 34 and 21 – 40. Multiply the opposite corners and subtract: $(18 \times 40) - (34 \times 21) = 720 - 714 = 6$, the triangular number at the head of the 18 – 21 column.

There are other relationships between the polygonal numbers. For example, $H_n = 4T_{n-1} + n$, where H_n is the *n*th hexagonal number, and T_n the *n*th triangular. Relationships such as this can be used to find the sum of the sequence of hexagonal numbers.

46

A famous, or infamous, example of numerology: in Psalm 46, the 46th word is 'shake'. The 46th word from the end counting backwards is 'spear'. Shakespear! Why? Well, when the King James Authorized Version was completed in 1610 (= 35 × 46), Shakespear was 46 years old!

47

With 48, a pair of Ulam numbers differing by 1. Muller calculated 20,000 terms and found no further such pairs. However, more than 60% of consecutive pairs differed by 2. [Muller: Guy 109]

$47 + 2 = 49$: $47 \times 2 = 94$. The pattern continues: $497 + 2 = 499$; $497 \times 2 = 994 \ldots$

48

The product of all the proper divisors of 48 is equal to 48^4.

If n is greater than 48, then there is a prime between n and $9n/8$, inclusive.

48 and 75 are the smallest pair of quasi-amicable or betrothed numbers. Each number is the sum of the proper divisors of the other, that is, the divisors excluding 1 and the number itself.

49

49 is trimorphic. Its cube ends in the same digits: $49^3 = 117,649$.

This is an example of a trimorphic number that is not automorphic.

$1/49 = 0 \cdot 02040\ 81632\ 65 \ldots$, in which the powers of 2 appear in sequence, eventually overlapping so that the pattern, although still there, cannot be seen.

49 is the first composite number with the property that all the fractions $n/49$, provided that n is not a multiple of 7, have periods that are cyclic permutations of each other. There are 42 such fractions, all of period 42. This occurs if and only if the number is a power of a prime whose reciprocal has maximum period. In this case $49 = 7^2$ and $1/7$ has period 6.

From 49 to 53 is the first quintuplet of consecutive numbers which are never values of $\sigma(n)$. The next two such quintuplets start at 115 and 145.

A square whose digits can be separated to make two other squares: $4/9$. So are $13^2 = 169$; $19^2 = 361$; $35^2 = 1225$; $38^2 = 1444$; $57^2 = 3249$; and $223^2 = 49,729$.

50

Denoted by the letter L in Roman numerals. The Romans had separate letters for 1, 10, 100, 1000, and for 5, 50 and 500.

The letter V stood for 5, and is often conjectured to represent 1 hand of 5 fingers, in which case the X for 10 could be 2 hands, or alternatively it could be an abbreviation for a row of 1s with a line through them to show that the round number 10 had been reached.

100 was C, the first letter of *centum*, and 1000 was M, the initial letter of *mille*. In between came D for 500.

By using these intermediate letters it was less tiresome and took less space to write numbers such as 856 or DCCCLVI, which would otherwise be the monstrous CCCCCCCCCXXXXXIIIIII.

50 is the smallest number to be the sum of 2 non-zero squares in 2 different ways. $50 = 5^2 + 5^2 = 7^2 + 1^2$. 65 is the smallest number for which all 4 squares are distinct and non-zero.

This follows from the fact that $50 = 5 \times 10$, the 2 smallest numbers that are each the sum of 2 squares.

In a passage in his *Republic* Plato refers to the 'rational diagonal of 5' meaning 7, which, because $7^2 = 50 - 1$, is very close to being the square root of 50, and the diagonal of a square of side 5. Plato knew that the actual diameter, $5\sqrt{2}$, is irrational.

The sequence of numbers that are sums of squares in two ways continues: 50 65 85 125 130 145 . . .

50 (followed by 52, 54 and 58) is the first number that can be partitioned into not more than 4 squares in 5 ways. [Sloane 53]

51

This appears to be the first uninteresting number, which of course makes it an especially interesting number, because it is the smallest number to have the property of being uninteresting.

It is therefore also the first number to be simultaneously interesting and uninteresting.

52

The number of weeks in a year, divided into 4 quarters of 13 weeks each, and also the number of cards in a standard pack without jokers, divided into 4 suits of 13 cards each.

The 3rd untouchable number, meaning that it is never a value of $\sigma(n) - n$. The sequence of untouchable numbers goes: 2 5 52 88 96 120 124 146 162 188 206 210 216 . . . [Sloane 1552]

The length of God's Algorithm for Sam Lloyd's Fifteen Puzzle (the maximum number of moves required from the worst possible initial position, with perfect play) is 52.

53

The smallest prime such that the period of its reciprocal is one-quarter of the maximum length possible, in this case, one-quarter of $53 - 1$, or 13. All fractions $k/53$, where k is a number between 1 and 52, fall into 4 classes, the decimal period of every fraction in the same class being a cyclic permutation of the others in its class.

Every positive integer is the sum of at most 53 4th powers.

If there are 53 people in a room, the chance that no pair of them have the same birthday is approximately 1/53: to be more precise, 1 in 53·01697 . . . [ApSimon, *More Mathematical Byways*, 1990]

The smallest prime which is not the difference between powers of 2 and 3. The sequence of such primes continues: 71 103 107 109 149 151 . . . [Sloane 5307]

55

The 10th triangular number: $55 = \frac{1}{2} \times 10 \times 11$.

It is also Fibonacci. The only Fibonacci numbers that are triangular are 0, 1, 3, 21 and 55.

55, 66 and 666 are the only triangular numbers that are composed of a repeated digit.

The sequence of triangular numbers that are palindromic starts: 1   3 6 55 66 171 595 666 3003 5995 8778 15,051 66,066 617,716 828,828 1,269,621 1,680,861 3,544,453 . . . [JRM v6]

Pyramidal numbers

The 5th square pyramidal number. If cannonballs are piled so that each layer is a square, then the total numbers of balls in successive piles will be 1, 5, 14, 30, 55, 91, 140 . . . The general formula for the *n*th number in the sequence is $\frac{1}{6}n(n + 1)(2n + 1)$.

Further pyramidal numbers can be defined by imagining that balls are being piled in pentagonal, hexagonal layers, and so on, but it is no longer possible actually physically to pile the balls up in a regular pattern. The formula for the number of balls in the *n*th 'pentagonal pyramid' is especially simple: $\frac{1}{2}n^2(n + 1)$.

The only numbers that are simultaneously triangular and square pyramidal are 1, 55, 91 and 208,335.

55 is the 4th Kaprekar number.

55 is a cubic recurring digital invariant. Add the cubes of its digits together; repeat twice, and 55 appears again:

$$55: \quad 5^3 + 5^3 = 250: \quad 2^3 + 5^3 + 0^3 = 133: \quad 1^3 + 3^3 + 3^3 = 55$$

Every number greater than 55 is the sum of distinct primes of the form $4n + 3$.

There are only 55 sets of integers *a*, *b*, *c*, *d* for which it is true that every integer is of the form $ax^2 + by^2 + cz^2 + du^2$. [Hardy, *Collected Papers of S. Ramanujan*, CUP, 1927]

55 multiplied by any of the odd numbers from 91 to 109 inclusive produces a palindrome.

Every even number n whose *abundancy index*, defined to equal $\sigma(n)/n$, is greater than 9, has at least 55 distinct prime factors. [Laatsch, MM v59 87]

56
Tetrahedral numbers
The 6th tetrahedral number. The sequence is: 1 4 10 20 35 56 84 120 ... with general formula: $\frac{1}{6}n(n + 1)(n + 2)$.

The traditional example of these numbers is a pile of cannonballs. The number of balls in each layer is, from the top downwards, 1, 3, 6, 10, 15 ... which is the sequence of triangular numbers, quite naturally, because each layer is triangular in shape. So the tetrahedral numbers can be thought of as the sums of the triangular numbers. Continuing into higher dimensions, in 4-dimensional space, the piles of tetrahedral numbers can themselves be piled up into 4-dimensional 'tetrahedrons', forming the 4-dimensional 'tetrahedral' numbers: 1 5 15 35 70 ... whose general formula is: $\frac{1}{24}n(n + 1)(n + 2)(n + 3)$.

The tetrahedral numbers can also be written as sums of squares: 1^2, 2^2, $1^2 + 3^2$, $2^2 + 4^2$, $1^2 + 3^2 + 5^2$ ...

57·296 ...
Approximately 57 degrees and 18 minutes. The number of degrees in 1 radian.

59
Euler posed the problem in 1772: to find a number that is the sum of 2 4th powers in 2 ways. He also found the smallest solution: $59^4 + 158^4 = 133^4 + 134^4$.

60
The base of a sexagesimal system of counting.

The Sumerians as early as 3500 BC had a decimal system for business purposes and a sexagesimal system used by a small number of experts, based on 10s and 6s: 1, 10, 60, 600, 3600, 36,000 ...

The Babylonians used this sexagesimal system for mathematical and astronomical work.

Systems based on 60 benefit from the many factors of 60. They have the advantages of a duodecimal system, and more.

In astronomy, the very ancient division of the Zodiac into 12 parts fits a sexagesimal system very well, and does not fit a decimal system at all.

The division of the circle into 360 degrees, and the division of degrees into 60 and 3600 parts originated among Babylonian astronomers a few centuries BC.

We still divide an hour of time or an angle of one degree into 60 minutes and each minute into 60 seconds. These are the only common measurements that have not been metricated.

60 degrees is the interior angle of an equilateral triangle.

60 is the smallest number which is the sum of 2 odd primes in 6 ways. The smallest numbers which are the sums of 2 odd primes in n ways, are: 6, 10, 22, 34, 48, 60 . . . [Sloane 4085]

Highly composite numbers
The 8th 'highly composite' number, defined by Ramanujan as a number that, counting from 1, sets a record for the number of its divisors. $60 = 2^2 \times 3 \times 5$ is the first number with 12 divisors.

The sequence of 'highly composite' numbers starts: 2 4 6 12 24 36 48 60 120 180 240 360 720 840 1260 1680 2520 5040 . . .

[Hardy, *Collected Papers of S. Ramanujan*, CUP, 1927]

61
$1,318,820,881^2 = 1,739,288,516,161,616,161$. (*See also 21.*)

$1/61$ has decimal period 60 which includes 6 occurrences of each of the digits 0 to 9, the smallest reciprocal whose period has this property. The next is $1/131$.

61, 62, 63, 64 is the first sequence of 4 consecutive integers with rising values of $\sigma(n)$, 62, 96, 104, 127, respectively. The next is 73, 74, 75, 76 with values 74, 114, 124, 140.

If a number of n digits is equal to the sum of the nth powers of its digits (making it a pluperfect digital invariant), then $n < 61$.

The smallest solution of the Pellian equation $x^2 - 61y^2 = 1$ is the large $x = 1,766,319,049$, $y = 226,153,980$.

62
There is no number which is 62 times the sum of its digits. Neither are there numbers which are 63, 65, 75, 84 . . . times the sum of their digits. [Conway, Sloane 5325]

63
Kaprekar's process for 2-digit numbers leads to the cycle: $63-27-45-9-81-63$ . . . In this cycle, 9 must be read as the 2-digit number 09.

For example, starting with 5 and 3: $53 - 35 = 18$; $81 - 18 = 63$, entering the cycle. Or, starting with 9 and 3: $93 - 39 = 54$; $54 - 45 = 9$, entering the cycle at a different point.

64

The second 6th power, after 1, and also a square and a cube: $64 = 4^3 = 8^2 = 2^6$.

It is therefore represented by 100 in octal and by 1,000,000 in binary.

The smallest number with 6 prime factors. The next smallest are 96, 128 (which has 7) and 144.

Being a cube, it is the sum of consecutive centred hexagonal numbers: $1 + 7 + 19 + 37 = 64$.

Fermat's Little Theorem says that if p is prime then $a^{(p-1)} - 1$ is divisible by p, provided a is not divisible by p.

For every prime p, there are values of a such that $a^{(p-1)} - 1$ is actually divisible by p^2.

The smallest such value for $p = 3$ is $8^2 = 64$: $64 - 1$ is divisible by $3^2 = 9$.

For $p = 5$, the next prime, the smallest solution is $7^4 - 1$, which is divisible by 25.

65

The second number to be the sum of two squares in two ways: $65 = 8^2 + 1^2 = 7^2 + 4^2$ and the first such to be the sum of two cubes: $4^3 + 1^3$.

65 is the magic constant in a 5 by 5 magic square.

66

The sum of the divisors of 66, including 66 itself, is a square: $1 + 2 + 3 + 6 + 11 + 22 + 33 + 66 = 144 = 12^2$.

The sequence of numbers with this property starts: 3 22 66 70 81 . . .

66, 105, 105 are a triple of triangular numbers, whose sums in pairs and whose totals are also triangular. [Ashbacher, Guy 189]

68

Any even integer greater than 68 can be written in at least 2 ways as the sum of 2 composite odd integers. [Ruemmler, and Minnich, MM v63 276]

69

The only number whose square and cube between them use all the digits 0 to 9 once each: $69^2 = 4761$ and $69^3 = 328,509$.

70

The sum of its divisors, including 70 itself, is a square, 144.

Weird numbers

The smallest weird number. A number is called weird if it is abundant without being the sum of any set of its own divisors. The factors of 70 are 1, 2, 5, 7, 10, 14 and 35, which sum to 74, so it is abundant, but no set of them sum to 70.

Weird numbers are rare. The only ones below 10,000 are 70, 836, 4030, 5830, 7192, 7912 and 9272.

Note that they are all even. It is not known whether an odd weird number exists. Professor Paul Erdös, who has the charming habit of offering money for the solutions to mathematical challenges, was offering, in 1971, $10 for the first example of an odd weird number, or $25 for a proof that none exist. This shows a nice judgement of the relative value of a counter example and a proof!

71

$71^2 = 7! + 1!$

$71^3 = 357,911$. The digits are the odd numbers 3 to 11 in sequence. [Davies, JRM v13]

The numbers 2, 5, 71, 369,119 and 415,074,643 are the only known numbers that divide the sum of all the primes less than them. [Sloane 1554]

72

$\phi(72) = \phi(78) = \phi(84) = \phi(90) = 24$. This is the smallest set of 4 numbers in arithmetical progression whose ϕ values are equal. The next two 4-term arithmetical progressions with equal ϕ values start at 216 and 76,236 and each also has common difference, 6. [Lal and Gillard, MOC v26]

There are 17 numbers n, a record up to that point, for which $\phi(n) = 72$. $\phi(n) = 96$ and $\phi(n) = 120$ also have 17 solutions.

The product of the number of edges (6), edges per face (3) and faces (4) of the tetrahedron. For the cube and octahedron, the same product is $4 \times 72 = 288$, and for the dodecahedron and icosahedron, $25 \times 72 = 1800$.

$72^5 = 19^5 + 43^5 + 46^5 + 47^5 + 67^5$ is the smallest 5th power equal to the sum of 5 other 5th powers.

73

All integers can be represented as the sum of at most 73 6th powers.

74

74 and 76 are the first pair of consecutive even numbers which are neither ever the values of $\phi(n)$.

76

$76^2 = 5776$, which ends in the digits 76, which is therefore called automorphic. The only other 2-digit automorphic number below 100 is 25. Automorphic numbers are related to multiples of powers of 10. For example, $76 \times 75 = 57 \times 10^2$.

It is not known whether there are any solutions in integers to the equation $76 = x^3 + y^3 + 2z^3$. The only other integers less than 1000 whose representation in this form is in doubt are 148, 183, 230, 356, 418, 428, 445, 482, 491, 580, 671, 788, 931 and 967. [Guy 151]

77

The smallest number in English requiring five syllables for its expression.

Every number greater than 77 is the sum of integers, the sum of whose reciprocals is 1. For example, $78 = 2 + 6 + 8 + 10 + 12 + 40$ and $1/2 + 1/6 + 1/8 + 1/10 + 1/12 + 1/40 = 1$. [Graham, *Journal of the Australian Mathematical Society*, 1963]

79

The smallest number that cannot be represented by less than 19 4th powers: $79 = 15 \times 1^4 + 4 \times 2^4$.

81

$81 = 3^4$

The sum of the divisors of 81 is 121, a square.

The fraction $1/81 = 0\cdot012345679\ 012345679\ 012 \ldots$ This pattern occurs because $81 = 9^2$ and 9 is 1 less than 10, the base of the decimal system.

In another base, 6 for example, the reciprocal of $(6 - 1)^2$ is $1/41 = 0\cdot01235\ 01235\ 01235 \ldots$

81 is the only number whose square root is equal to the sum of its digits, apart from the trivial 0 and 1.

81 is both square and heptagonal.

The heptagonal numbers, with formula $n(5n - 3)/2$, form the sequence: 1 7 18 34 55 81 112 148 189 235 286 . . .

It is the smallest square whose sum of divisors is also square. $\sigma(81) = 11^2$. The next such example is $\sigma(20^2) = 31^2$. [JRM v27 227]

Write the natural numbers in groups, like this:

$$1 \quad 2,3 \quad 4,5,6 \quad 7,8,9,10 \quad 11,12,13,14,15 \quad \ldots$$

Delete every second group. The sum of the first remaining n groups is then n^4. [Juzuk, *Scripta Mathematica*, 1939] For example,

$$1 + 4 + 5 + 6 + 11 + 12 + 13 + 14 + 15 = 81 = 3^4$$

More straightforward is this pattern:

$$1 = 3^0$$
$$2 + 3 + 4 = 3^2$$
$$5 + 6 + 7 + 8 + 9 + 10 + 11 + 12 + 13 = 3^4$$
$$14 + 15 + 16 + \ldots + 39 + 40 = 3^6$$

and so on. [Khatri, *Scripta Mathematica* v20] The number of terms in each sequence is 1, 3, 9, 27 . . .

84

Little is known of the life of Diophantus. This verse from *The Greek Anthology* purports to give his age, which turns out to be 84.

This tomb holds Diophantus. Ah, how great a marvel! The tomb tells scientifically the measure of his life. God granted him to be a boy for one-sixth of his life, and adding a twelfth part to this, he clothed his cheeks with down. He lit him the light of wedlock after a seventh part, and five years after his marriage he gave him a son. Alas, late-born wretched child! After attaining the measure of half his father's life, chill Fate took him. After consoling his grief by the study of numbers for four years, Diophantus ended his life.

85

The sum of 2 squares in 2 ways: $85 = 9^2 + 2^2 = 7^2 + 6^2$.

86

2^{86} in base 10 contains no zero. This is probably the largest such power of 2. [Sloane 485]

87

$\sigma(\phi(87)) = \sigma(87)$ Other solutions, apart from 1, are 362, 1257, 1798, 5002, 9374. [Guy 99]

88

88 is itself a repeated digit, and its square consists of pairs of repeated digits $88^2 = 7744$.

88, 89, 90 are three successive numbers none of which is the sum of three pentagonal numbers. Other such triplets are 19, 20, 21; 99, 100, 101; and 111, 112, 113. [AMM v101 171]

89

89 and 97 are the first pair of consecutive primes differing by 8.

Double 89 and add 1: repeat, to get a sequence of 6 Sophie Germain primes, 89   179   359   719   1439   2879.

This is the smallest such 6-prime sequence. [JRM v13]

Add the squares of the digits of any number: repeat this process, and eventually the number either sticks at 1, or goes round this cycle: 89–145–42–20–4–16–37–58–89 . . .

89 and 98 are the 2-digit numbers that require most reversals-and-adding to become palindromes. They each require 24 steps.

89 is the 11th Fibonacci number, and the 5th Fibonacci prime, and the period of its reciprocal is generated by the Fibonacci sequence: $1/89 = 0·01123\ 5$ . . . (because $89 = 10^2 - 10 - 1$).

90

The number of degrees in a right angle.

91

The number of days in a quarter-year, counted as 13 weeks of 7 days each.

91 is simultaneously triangular, equal to $1 + 2 + 3 + . . . + 13$; square pyramidal, equal to $1^2 + 2^2 + . . . + 6^2$; and a centred hexagonal number equal to $1 + 6 + 12 + 18 + 24 + 30$.

91 is the smallest pseudoprime to base 3. That is, $3^{90} - 1$ is divisible by 91 although 91 is not a prime but 7×13.

The sequence of pseudoprimes to base 3 continues: 121, 286, 671, 703 . . .

91 is pseudoprime to 35 bases less than itself. [Tiger Redman]

94

The smallest even number apart from 2 and 4 that is not the sum of 2 of the sequence of twin primes: 3   5   7   11   13   17   19   29   31   41   43 . . .

96 and 98 are also not the sum of 2 twin primes. The next numbers to fail are 400, 402 and 404.

96

The second smallest number with 6 prime factors: $96 = 3 \times 2 \times 2 \times 2 \times 2 \times 2$.

97

The period of its decimal reciprocal is a maximum, of length 96. Alexander Aitken, a lightning calculator who was also a professor of mathematics at Edinburgh University, knew it off by heart. He will hardly have been helped significantly by the fact that it starts with the powers of 3 (because $97 = 100 - 3$):

$1/97 = 0.01030$ 92783 50515 46391 75257 73195 87628 86597 93814 43298 96907 21649 48453 60824 74226 80412 37113 40206 18556 7 . . .

97·40909 10340 0 . . .
π^4.

The continued fraction for π^4 starts:

$$97 + \cfrac{1}{2 + \cfrac{1}{2 + \cfrac{1}{3 + \cfrac{1}{1 + \cfrac{1}{16539}}}}}$$

Truncating just before the unexpectedly large partial quotient 16,539 gives a famous approximation of Ramanujan for π^4 of $97\frac{9}{22}$.

98

The period of its decimal reciprocal starts with the powers of 2: $1/98 = 0.01020$ 40816 32653 06122 44897 95918 36734 69387 75510 204 . . .

The smallest number which cannot be written as the sum of two odd primes with one of 3, 5, 7 as the smaller. The next such numbers are 122, 124, 126 and 128. The minimum-smaller-prime representations are $98 = 19 + 79$, $122 = 13 + 109$, $124 = 11 + 113$, $126 = 13 + 113$ and $128 = 19 + 109$. [Sloane 2273]

99

$1/99 = 0.01$ 01 01 . . .

9 and 11 have very simple reciprocals as decimals, because $9 \times 11 = 99$.

Similarly, $27 \times 37 = 999$.

99 is a Kaprekar number, as is any string of 9s. $99^2 = 9801$ and $98 + 01 = 99$.

100

The square of 10, the base of the decimal system, but also the square of the base in any other base.

The boiling point of water on the Celsius (Centigrade) scale of temperature.

Denoted by C by the Romans, from *centum* meaning hundred.

In the metric system, the prefix 'centi' means one-hundredth, as in centimetre, one-hundredth of a metre.

Because 10 is the 4th triangular number, $100 = 10^2$ is the sum of the first 4 cubes: $100 = 1^3 + 2^3 + 3^3 + 4^3$.

It is a very old puzzle to join the digits 1 to 9, in that order, using only the usual signs of operations, and brackets, to make a total of 100.

Dudeney gives many solutions, including this one, which he describes as the usual answer: $1 + 2 + 3 + 4 + 5 + 6 + 7 + (8 \times 9) = 100$.

His own preferred solution, because it requires the use of only 3 signs, is: $123 - 45 - 67 + 89 = 100$.

Alternatively, using the digits in reverse order, $98 - 76 + 54 + 3 + 21 = 100$.

The largest number for which the sum of the primes less than the number of primes less than or equal to the number is the number itself. In this case, $\pi(n) = 25$, and the sum of the primes from 2 to 23 = 100. The other numbers with this property are 5, 17, 41 and 77. [Golomb, AMM v98 858]

101

It is not known if there is an infinite number of palindromic primes. 101 is the smallest, apart from the 1-digit primes, 2, 3, 5 and 7, and 11.

The other palindromic primes below 1000 are 131, 151, 181, 191, 313, 353, 373, 383, 727, 757, 787, 797, 919, 929.

102

$102^7 = 12^7 + 35^7 + 53^7 + 58^7 + 64^7 + 83^7 + 85^7 + 90^7$

102^7 is the smallest 7th power to be the sum of only 8 other 7th powers.

103

103 is the smallest prime the period of whose reciprocal is one-third of the maximum length. $1/103$ has period of length 34. One-third of all the fractions $n/103$ where n is less than 103 have periods that are cyclic permutations of this. The other two-thirds share 2 different 34 digit periods.

104

104 is semi-perfect, because it is the sum of some of its own divisors: $104 = 52 + 26 + 13 + 8 + 4 + 1$.

It is irreducibly semi-perfect because no factor of 104 is itself semi-perfect.

$\phi(104) = \phi(105) = 48$. The sequence of n such that $\phi(n) = \phi(n + 1)$ goes: 1   3   15   104   164   194   255   495   584   975   2204 ...

105

Erdös conjectured that this is the largest number n such that the positive values of $n - 2^k$ are all prime. The only other known numbers with this property are 7, 15, 21, 45 and 75. [Guy 42]

105 is the largest integer such that every odd integer less than it and prime to it is a prime. [Golomb, AMM v94 883]

$\phi(105) = 48$, $\phi(106) = 52$, $\phi(107) = 106$, the first sequence of three successive strictly increasing values of $\phi(n)$. The next two such sets are $\phi(165) = 80$, $\phi(166) = 82$, $\phi(167) = 166$, and $\phi(315) = 144$, $\phi(316) = 156$, $\phi(317) = 316$.

105 is the second number n such that $\phi(n) \times \nu(n) = \sigma(n)$, where $\nu(n)$ is the number of divisors of n. $\phi(105) = 48$, $\nu(105) = 8$ and $\sigma(105) = 192$.

The first such number is 35.

105 is the smallest number n, such that 1 can be represented as a sum of odd reciprocals, none of them less than $1/n$: $1 = 1/3 + 1/5 + 1/7 + 1/9 + 1/11 + 1/33 + 1/35 + 1/45 + 1/55 + 1/77 + 1/105$.

There are 4 ways of representing 1 as the sum of odd reciprocals, using only 9 of them, but in each case the smallest is less than 1/105. The solution with the largest least term is: $1 = 1/3 + 1/5 + 1/7 + 1/9 + 1/11 + 1/15 + 1/35 + 1/45 + 1/231$. [Kierstead and Nelson, JRM, v10]

108

There are 108 heptominoes, one of which surrounds a hole. It is in the form of a 3×3 square with one corner and the central square missing. This is the smallest polyomino to contain a hole.

The number of its factors is also a cube, 8.

110

$\sigma(110) = 216 = 6^3$, the first n, after $n = 1$, 7, for which $\sigma(n)$ is a cube.

111

The magic constant for the smallest magic square composed only of prime numbers, counting 1 as a prime.

The second repunit, composed only of the digit 1.

A palindromic Lucky number. The sequence of these runs: 1 3 7 9 33 99 111 141 151 171 . . .

A sequence of 111 consecutive composite numbers runs from 370,262 to 370,372, the first run of more than 100. [MG v78 168]

$111 = 20^2 - 17^2$, the third difference of 2 squares equal to a repunit. The sequence of such squares starts: 1, 0; 6, 5; 20, 17; 56, 45; 156, 115; 344, 85; 356, 125 . . . [Lacampagne and Selfridge, MM v59 270]

112

There are $112 = 4 \times 28$ pounds in a hundredweight.

The length of the side of the smallest possible dissection of a square into 21 other distinct squares.

The side of the smallest equilateral triangle such that there is an interior point at integral distances from all three vertices. These distances are 57, 65 and 73. [ApSimon, *More Mathematical Byways*, 1990]

113

The smallest 3-digit prime such that all other arrangements of its digits are also prime numbers.

Other such prime numbers are 337 and 199, and their rearrangements. The 2-digit primes with this property are 11, 13, 17, 37 and 79.

1,111,111,111,111,111,111 and 11,111,111,111,111,111,111 are the next two numbers with the property. [MM v47 233]

The first prime gap as large as 14 occurs between 113 and 127. The next two such occur from 293 to 307, and from 317 to 331. The next record-breaking gap is between 523 and 541.

114

The number of ways of colouring the faces of a cube with 3 different colours. With 4 colours, there are 2652 patterns and with 5 colours, 29,660.

118

118 is the smallest number which can be written as the sum of 4 triples, whose products are all equal: $118 = 14 + 50 + 54 = 15 + 40 + 63 = 18 + 30 + 70 = 21 + 25 + 72$.

The product of each triple is 37,800. [Mauldron, Guy 172]

If n is greater than or equal to 118, then the interval n to $4n/3$ inclusive contains a prime number of each of the forms $4n + 1$, $4n - 1$, $6n + 1$ and $6n - 1$.

The sum of the cubes of the 5 consecutive positive integers 118 to 122 is a square. The only other similar sequences with this property start with 1 and 96.

120

$120 = 1 \times 2 \times 3 \times 4 \times 5 = 5! = 4 \times 5 \times 6$. It is also 4! in base 4.

120 is also the 15th triangular number and the 8th tetrahedral number, formed by summing the triangular numbers: $120 = 1 + 3 + 6 + 10 + \ldots + 28 + 36$.

120 is the smallest number to appear 6 times in Pascal's triangle.

It is the smallest multiple of 6, such that $6n + 1$ and $6n - 1$ are both composite.

120 is the smallest number having $16 = 2^4$ divisors.

The product of any two of the numbers 1, 3, 8 and 120, plus 1, is a square. Baker and Davenport proved that a 5th number cannot be added to the set so that this property is preserved.

Multiply-perfect numbers

The ubiquitous Marin Mersenne discovered that the factors of 120 sum to $2 \times 120 = 240$, and proposed to his friend Descartes the problem of finding further numbers whose factors sum to a multiple of the original number.

$120 = 2^3 \times 3 \times 5$ and its factors, 1, 2, 3, 4, 5, 6, 8, 10, 12, 15, 20, 24, 30, 40 and 60 sum to $240 = 2 \times 120$.

If 120 is counted among its own factors, then the sum is $360 = 3 \times 120$ and for this reason 120 is sometimes called tri-perfect, or multiply perfect of order 3, in which case ordinary perfect numbers are of order 2.

Only 6 tri-perfect numbers are known:

120, 672, 523,776, 459,818,240, 1,476,304,896, 31,001,180,160.

These are all even, just as all known perfect numbers are even. If an odd tri-perfect number exists, then it exceeds 10^{70}, and has at least 11 distinct prime factors. If it is not divisible by 3, then it is even larger, greater than 10^{108} with at least 32 distinct prime factors.

Hundreds of multiply-perfect numbers are known, of order up to 9. One of the smallest of order 8 was discovered by Alan L. Brown, an American 'human computer': $2 \times 3^{23} \times 5^9 \times 7^{12} \times 11^3 \times 13^3 \times 17^2 \times 19^2 \times 23 \times 29^2 \times 31^2 \times 37 \times 41 \times 53 \times 61 \times 67^2 \times 71^2 \times 73 \times 83 \times 89 \times 103 \times 127 \times 131 \times 149 \times 211 \times 307 \times 331 \times 463 \times 521 \times 683 \times 709 \times 1279 \times 2141 \times 2557 \times 5113 \times 6481 \times 10,429 \times 20,857 \times 110,563 \times 599,479 \times 16,148,168,401$. [Beck and Najar, MOC, 1982]

121

A palindromic square of a palindrome, and a perfect square in any base from 3 upwards.

$11^3 = 1331$ and $11^4 = 14,641$ are also palindromic.

Brocard's problem: are $4! + 1 = 5^2$, $5! + 1 = 11^2$ and $7! + 1 = 71^2$ are the only equations of this form?

Fermat conjectured correctly that 121 and 4 are the only squares that become cubes when increased by 4.

121 is the only square that is the sum of consecutive powers from 1: $121 = 1 + 3 + 9 + 27 + 81$.

Every number greater than 121 is the sum of distinct primes of the form $4n + 1$.

11^2 in base 10, and also 11 in factorial base, because $11 = 3! + 2 \times 2! + 1!$

A palindrome in base 10 and also in base 3: $121 = 11,111$ (base 3). The sequence of palindromes to base 3 and 10 goes: 1 2 4 8 121 151 212 242 484 656 757 . . . [JRM v18 169]

The second Wonderful Demlo number. The sequence goes: 1 121 12,321 1,234,321 123,454,321 . . . The sequence is infinite because 9 is followed by 0. In case this seems completely trivial, note that these numbers are the coefficients in the expansion of:
$$(1 + 10x)/(1 - x)(1 - 10x)(1 - 100x)$$

125

A cube, 5^3, which is the sum of two squares in two ways: $125 = 10^2 + 5^2 = 11^2 + 2^2$.

$5^3 - 2^7 = 5 - 2$. The only known matching pattern is $13^3 - 3^7 = 13 - 3$. [Guy 155]

125 is the decimal part of $1/8$. Because $8 = 10 - 2$, it can be written as a sum, similar to the sums that are related to periodic decimals:

```
1 2 4 8
    1 6
      3 2
        6 4
          1 2 8
            2 5 6
              5 1 2
                . . . .
1 2 4 9 9 9 9 9 9 9 9 . . .
```

126

The area of the third-smallest triangle with integral sides in arithmetical progression, and integral area: the sides are 15, 28, 41. The first two are 3, 4, 5 and 13, 14, 15; the next is 15, 26, 37, with area 156. [MacNeill, *Mathematical Spectrum* 21–3]

127

In 1848 de Polignac conjectured that every odd number could be expressed as the sum of a power of 2 and a prime. He claimed verification up to 3 million, but 127 fails, for starters. [Boston, *Quarch*, no. 6] Other numbers which fail, are 149, 251, 331, 337 . . . [Sloane 5390]

Mersenne numbers

$127 = 2^7 - 1$ is the 7th Mersenne number, denoted by M_7, and the 4th Mersenne prime, and therefore the source of the 4th perfect number.

Father Marin Mersenne was a natural philosopher, theologian, mathematician and musical theorist, and the moving spirit of one of the most important French scientific groups of the early seventeenth century. He was a friend of Descartes, with whom he studied at a Jesuit college, Desargues, Fermat, Frenicle and the Pascals, father and son, and other mathematicians, to whom he proposed problems concerning perfect numbers and related ideas.

In 1644, in the Preface to *Cogitata Physico-Mathematica* he asserted that the only values of p not greater than 257 for which $2^p - 1$ is prime are 1, 2, 3, 5, 7, 13, 17, 19, 31, 67, 127 and 257. Mersenne counted 1 as prime. Nowadays the list starts with M_2.

The first four of these, which are 3, 7, 31 and 127, are obviously prime.

M_{13} was known to be prime in medieval times and M_{17} and M_{19} were also known to be prime. So Mersenne was stating that between 31 and 257 inclusive there are only 4 prime M_p: M_{31}, M_{67}, M_{127} and M_{257}. (Mersenne knew that M_p must be composite if p is composite, but the converse is not true.) Mersenne was effectively making a statement about all the prime powers of 2 up to and including 2^{257}.

A most remarkable claim bearing in mind his complete lack of modern computers, and the size of the larger numbers. Fermat already knew (1640) that any factor of the Mersenne number M_p must be of the form $2np + 1$, but this fails to eliminate an enormous number of possible large factors. Perhaps he was relying on Fermat for some theorem or idea that is now lost. Anyway, the list contains mistakes, though these were all discovered long after Mersenne's death. In fact 2 out of 4 of Mersenne's additions to the list are actually composite, and he missed 3 primes.

M_{61} is prime, proved by Pervusin in 1883, and M_{67} is composite (67 as printed might possibly have been an error for 61); M_{89} and M_{107}, which Mersenne omitted, are both prime, and M_{257} is composite.

On the other hand, M_{31} is indeed prime, and so is the gigantic M_{127}.

Mersenne's list, despite or perhaps because it was so ambitious and erroneous, has provided a stimulus to mathematicians to invent better and better methods of solving one of the very simplest problems in mathematics, so simple indeed that any child who has learned to do long multiplication can understand it but which mathematicians can still only partially solve.

The problem is to reverse the result of multiplication, that is, to take a large number and decide whether it is the product of at least 2 other numbers, and if so to find them. So simple to state, so difficult to do!

Mersenne numbers are ideal candidates for even relatively elementary methods, because they are constructed in such a simple manner, just like the Fermat numbers $2^{2^n} + 1$. Both sets of numbers are strikingly non-random and their structure provides the basis for their factorization.

With the advent of modern computers, many far larger Mersenne primes have been discovered, starting with $2^{521} - 1$ in 1952, each leading to an even perfect number. All the Mersenne primes known to date are listed under *28: Perfect numbers*.

128

2^7 and therefore in binary 10,000,000.

The smallest number to be the product of 7 prime factors.

128 is a power of 2, all of whose digits are powers of 2. It is the only such up to $2^{70,000}$. [Saunders, JRM v26 151]

128 is the largest number that is not the sum of distinct squares. [Sprague]

132

132 is the sum of all the 2-digit numbers made from its digits: $132 = 13 + 32 + 21 + 31 + 23 + 12$.

It is the smallest such number.

133·335

The Dewey Decimal classification for 'numerology'. Martin Gardner, in *The Numerology of Dr Matrix*, points out that if you reverse it, and add: $133·335 + 533·331 = 666·666$, you discover the Number of the Beast, repeated! Deeply significant!

135

$135 = 1^1 + 3^2 + 5^3$.

Other examples of the same pattern are: $175 = 1^1 + 7^2 + 5^3$; $518 = 5^1 + 1^2 + 8^3$ and $598 = 5^1 + 9^2 + 8^3$.

136

Sum the cubes of its digits: $1^3 + 3^3 + 6^3 = 244$. Repeat and the original number returns: $2^3 + 4^3 + 4^3 = 136$.

139

139 and 149 are the first consecutive primes differing by 10.

140

With 195, the second pair of betrothed or quasi-amicable numbers. $\sigma(140) = \sigma(195) = 140 + 195 + 1$. The first pair is (48,75), the third (1575,1648). [Guy 59]

A number is harmonic if the harmonic mean of all its divisors is an integer. The perfect numbers are necessarily harmonic. 140 is the smallest harmonic number, apart from 1 and the perfect numbers.

The arithmetic mean of its divisors is also integral. The sequence of numbers with integral arithmetic and harmonic means starts: 1 6 140 270 672 . . .

141

Cullen numbers are of the form $n \times 2^n + 1$. Cullen numbers are prime for $n = 1, 141, 4713, 5795, 6611$ and $18,496$, and composite for every other $n \leqslant 30,000$. [Keller, MOC v64 1733]

On the other hand, numbers of the form $n \times 2^n - 1$ are prime 6 times below 100, for $n = 2, 3, 6, 30, 75$ and 81.

144

12^2, a gross or a dozen dozen, and therefore '100' in the duodecimal system of counting.

144 is the only square Fibonacci number, apart from 1. Moreover it is the 12th Fibonacci number.

A divisor of a Fibonacci number is called proper if it does not divide any smaller Fibonacci number. The only Fibonacci numbers that do not possess a proper divisor are 1, 8 and 144.

144 ends in a repeated '44'. A square can end in a repeated digit only if it is a multiple of 100, or if the root ends in 12, 38, 62 or 88, when the square ends in '44'.

Reversing 12 and 144 gives $441 = 21^2$.

The smallest magic square composed of consecutive primes comprises the 144 odd primes from 3 upwards. The magic constant is 4515.

Euler conjectured that no nth power can be the sum of fewer than n nth powers. For example, a cube cannot be the sum of only two cubes, which is true. (It is the smallest case of Fermat's Last Theorem.)

In 1966 L. J. Lander and T. R. Parkin were searching on computer for 5th powers that were the sum of 5 other 5th powers. To their great surprise they not only found 4 solutions to their original problem – but in one solution, one of the numbers was 0^5, so they had in fact discovered a counter-example to Euler's conjecture: $144^5 = 27^5 + 84^5 + 110^5 + 133^5$.

No other 5th power up to 765^5 can be expressed as the sum of only 4 5th powers, apart of course from the multiples of 144: 288, 432, 576 and 720.

145

$145 = 1! + 4! + 5!$

The only other numbers that are the sum of the factorials of their digits are 1, 2 and 40,585.

The 5th number to be the sum of 2 squares in 2 different ways: $145 = 12^2 + 1^2 = 8^2 + 9^2$.

147

The number of ways of representing 1 as the sum of 5 unit fractions, in decreasing order of size. The number of representations for 1, 2, 3 . . . unit fractions is 1, 1, 3, 14, 147, 3462 . . . [Singmaster, Guy 162]

153

$153 = 1! + 2! + 3! + 4! + 5!$

When the cubes of the digits of any number that is a multiple of 3 are added, and then this process is repeated, the final result is 153, where the process ends, because $153 = 1^3 + 5^3 + 3^3$. For all number less than 10^8 at most 14 cycles are needed to reach 153. [S. S. Gupta]

The other 3-digit numbers that equal the sum of the cubes of their own digits are 370, 371 and 407.

These pairs switch from one to the other in a 2-cycle: 136 and 244; 919 and 1459.

There are two cycles of length 3: 55–250–133 and 160–217–352.

When G. H. Hardy wished, in his book *A Mathematician's Apology*, to give examples of mathematical theorems that were not 'serious', he chose two examples, 'almost at random, from Rouse Ball's *Mathematical Recreations*'. The first was the fact that 8712 and 9801 are the only 4-digit numbers that are multiples of their reversals. The second was the fact that, apart from 1, there are just 4 numbers that are the sums of the cubes of their digits, those mentioned above. Hardy commented,

These are odd facts, very suitable for puzzle columns and likely to amuse amateurs, but there is nothing in them that appeals to the mathematician. The proofs are neither difficult nor interesting – merely a little tiresome. The theorems are not serious; and it is plain that one reason . . . is the extreme speciality of both the enunciations and the proofs, which are not capable of any significant generalization.

As any critic might have remarked of Euler's solution of the Bridges of Königsberg problem, or of Euler's dabbling in Magic Squares. The existence or non-existence of significant generalizations would appear to be a contingent fact, not susceptible to proof by G. H. Hardy.

As an almost certainly less interesting fact I would suggest, 'The 10,000,000 digit of π is a 7,' mentioned by Keith Devlin, though the supposed lack of interest in this fact is still a matter of contingent fact, and no more.

In the New Testament the net that Simon Peter drew from the sea of Tiberias held 153 fishes. This was inevitably interpreted numerologically by the early Church fathers, especially St Augustine.

153 is the 17th triangular number and therefore already significant. But what is special about 17 itself? It is the sum of 10 for the Ten Commandments of the Old Testament to 7, for the Gifts of the Spirit in the New Testament. This was a common means of combining two influences, just as the Pythagoreans associated 5 with marriage because $5 = 2 + 3$ and those numbers are female and male respectively.

W. E. Bowman, a modern writer with more humour and less reverence, introduces the number 153 on numerous occasions into his novel *The Ascent of Rum Doodle*. It appears as the height of the ship above sea level, the speed of a train chugging through the foothills of the Himalayas, the number of porters to be hired for the ascent, and the depth of a crevasse, among other things.

157

The smallest number for which $\phi(2n + 1) < \phi(2n)$. The sequence of such numbers continues: 262 367 412 . . .

The largest odd integer which cannot be expressed as the sum of 4 distinct non-zero squares with greatest common divisor 1. If 5 squares are allowed, under the same conditions, the largest non-representable number is 245. [*Acta Arithmetica* v67 349]

159

159 cannot be represented as the sum of fewer than 19 4th powers.

The largest number that is not the sum of distinct pentagonal numbers.

161

Every number greater than 161 is the sum of distinct primes of the form $6n - 1$.

163

Aitken competed successfully with Wim Klein, a Dutch prodigy who had memorized the multiplication table up to 100×100 but lacked the mathematical knowledge to employ clever short cuts. Aitken often made subconscious calculations. He told of results that 'came up from the murk', and would say of a particular number that it 'feels prime' as indeed it was. He was one of the few to whom integers were personal friends. He noticed, for instance, an amusing property of 163: that $e^{\pi\sqrt{163}}$ differs from an integer by less than 10^{-12}. As he himself once put it, 'Familiarity with numbers, acquired by innate faculty sharpened by assiduous practice, does give insight into the profounder theorems of algebra and analysis'. [Ball and Coxeter, *Mathematical Recreations and Essays*, 1974]

168

The product $nd(n)$ has the same value for each of the triplet $n = 168$, 192 or 224. There are three smaller pairs for which $nd(n)$ has the same values: 18 and 27; 24 and 32; 56 and 64. [Guy 68]

$\sigma(60) = 168$. This peak in the values of $\sigma(n)$ exceeds the previous peak ($\sigma(48) = 124$) by 44. The next larger difference between peaks occurs between $\sigma(108) = 280$ and $\sigma(120) = 360$.

There are 168 primes with at most 3 digits, from 2 to 997. The sequence of numbers of primes with at most n digits starts: 4 25 168 1229 9592 78,498 . . . [Sloane 3608]

169

$169 = 13^2$ and $961 = 31^2$

The smallest square hexagonal number, apart from 1. The next smallest are 32,761 and 6,355,441. [Sloane 5409]

169 is a square itself and can be expressed as the sum of either 2, 3 or 4 non-zero squares: $13^2 = 12^2 + 5^2 = 12^2 + 4^2 + 3^2 = 8^2 + 8^2 + 5^2 + 4^2$. In fact, 169 can be written as the sum of n non-zero squares, for all values of n from 1 to 155, but for no larger values. [Jackson, Masat and Mitchell, MM v61 41]

Adding the factorials of its digits: $1! + 6! + 9! = 363,601$. Repeat the process with 363,601 to get 1454. Repeat again to get back to 169.

The smallest square which is the difference between 2 cubes, 8^3 and 7^3.

175

$175 = 1^1 + 7^2 + 5^3$

180

The number of degrees in a half-circle, and the number of degrees Fahrenheit between the freezing point of water, 32, and its boiling point, 212.

The sum of the angles of a triangle.

180^3 is the sum of consecutive cubes: $180^3 = 6^3 + 7^3 + 8^3 + \ldots + 68^3 + 69^3$.

187

The smallest of a group of 3-digit numbers that require 23 reversals to form a palindrome.

188

All numbers greater than 188 can be expressed as the sum of at most 5 distinct squares. Only the numbers 188 and 124 require as many as 6. [Bohman, Fröberg and Riesel, Guy 136]

192

$192 + 384 = 576$, a pandigital sum, where $384 = 2 \times 192$ and $576 = 3 \times 192$. There are 3 other such sums, starting with 219, 273 and 327. [Singmaster, JRM v27 183]

196

$196 = 14^2$ has the same digits as $169 = 13^2$.

The 6th heptagonal pyramidal number, with formula $\frac{1}{6}n(n + 1)(5n - 2)$, and also a square.

Palindromes by reversal

If 87 is reversed and added to itself, and the process is repeated, then after only four steps it produces a palindrome, 4884: $87 + 78 = 165$: $165 + 561 = 726$: $726 + 627 = 1353$: $1353 + 3531 = 4884$.

This is effectively a statement about the size of the digits at the previous step. To obtain a palindrome it is sufficient that at the previous addition there should be no carries and therefore that the digits of the previous stage, taken in pairs from either end, should sum to 9 or less. Do all numbers become palindromes eventually? The answer to this problem is not known. 196 is the smallest number less than 10,000 that by this process has not yet produced a palindrome. John Walker and Tim Irvin have carried the calculation for 196 to 1,000,000 digits (reached after 2,415,836 reversals and additions, taking 3 years' spare time on a Sun 3/260) and then 2,000,000 digits (2 months' spare time, only, on a supercomputer) respectively, without finding a palindrome. [Pickover, *Computers and the Imagination*, 1991]

Of the 900 3-digit numbers, 90 are themselves palindromic, 735 require from 1 to 5 reversals only.

The remaining 75 numbers can be classed into just a few groups, the members of which after one or two reversals each produce the same number and are therefore essentially the same. One of these groups consists of the numbers 187, 286, 385, 583, 682, 781, 869, 880 and 968, each of which when reversed once or twice forms 1837 and eventually forms the palindromic number 8,813,200,023,188 after 23 reversals. [Richard Hamilton]

Among the first 100,000 numbers there are 5996 that have been found not to create a palindrome. Since the probability that a randomly chosen number will have digits that when paired from the ends always sum to 9 or less clearly decreases with the length of the number, it is plausible to suppose that the larger the number, the smaller the chance that a palindrome will ever appear.

In base 2, it is certainly not true that every number eventually generates a palindrome. Roland Sprague shows that 10110 never does so.

199
$199 + 210n$ for $n = 0, 1, 2, 3, 4, 5, 6, 7, 8, 9$ provides the smallest 8, 9 and 10 primes in arithmetical progression.

200
The smallest number which cannot be changed into a prime by changing one digit. With 202, 204, 206, 208 it forms an arithmetic progression of numbers with the same property. [Sierpinski, *250 Problems in Elementary Mathematics*, no. 101]

204
204^2 is the sum of consecutive cubes: $204^2 = 23^3 + 24^3 + 25^3$.

205
Every number greater than 205 is the sum of distinct primes of the form $6n + 1$.

210

$$7\# = 2 \times 3 \times 5 \times 7$$

210 is triangular and pentagonal. The smallest such number is, as usual, 1, and the next smallest is 40,755.

$\phi(210) = 48$ is a factor of $\sigma(210) = 576$, and $d(210) = 16$ divides both. The sequence of numbers with both these properties starts: 1 3 15 30 35 52 70 78 105 140 168 190 210 . . .

The number of representations of n as the sum of two primes is at most the number of primes in the interval $[n/2,\ n-2]$. 210 is the largest n for which the maximum is reached. [MOC v61 209]

212

The boiling point of water in degrees Fahrenheit.

216

$216 = 6^3$ is the smallest cube that is also the sum of 3 cubes: $216 = 3^3 + 4^3 + 5^3$. The next smallest is $9^3 = 1^3 + 6^3 + 8^3$.

This dissection can be demonstrated physically by dissecting a cube, using only 8 pieces.

216 is the magic constant in the smallest possible multiplicative magic square, discovered by Dudeney.

Plato's number

The famous and notorious number of Plato occurs in an obscure passage in *The Republic*, viii, 546 B–D, which starts,

> But the number of a human creature is the first number in which root and square increases, having received three distances and four limits, of elements that make both like and unlike and wax and wane, render all things conversable and rational with one another.

This is merely the beginning of the passage. It illustrates perfectly both the intimate relationship that Plato, as a Pythagorean, perceived between numbers and the real world, and the difficulty that he had in using the then available language to express himself. Mathematical language was not well developed in Plato's time, and so he often apparently called upon the resources of everyday language. I say 'apparently' because some words in the passage are hardly known in other preserved writings and therefore their meaning is especially difficult to interpret. (The obscurity is not entirely due to our distance from Plato in time. Early Greek commentators also found the passage difficult.)

The whole passage has been analysed in the minutest detail by innumerable commentators. Two numbers are actually involved and the smaller it is agreed is 216, though this is variously derived. (The larger is 12,960,000.)

The well-known 3 – 4 – 5 Pythagorean triangle has area 6. The expression 'three distances and four limits' is supposed to refer to cubing. Adams eventually reaches the conclusion that the number intended in the quoted passage is 216 as the sum of the cubes of the sides of the triangle. However, it has also been deduced as the cube of 2×3.

2 and 3 were associated with female and male respectively, and 5 with marriage. 6 also was associated with marriage, being 2×3 rather than $2 + 3$. Given the Pythagoreans' basic belief in the efficacy of numbers in interpreting the world, it can hardly be denied that such number-theoretic relationships as this support their approach.

[Adams, *The Republic of Plato*, CUP, 1929]

219

There are 219 space groups in 3 dimensions. They are the analogues of the 17 basic wallpaper patterns in 2 dimensions, and determine the possible shapes of mineral crystals.

11 of them however come in 2 forms, with a left-hand screw or a right-hand screw. This difference is important in the structure and optical properties of crystals, so from this point of view there are 230 space groups.

220

Amicable numbers

220 and 284 form the first and smallest amicable pair. Each is the sum of the aliquot parts of the other: $220 = 2^2 \times 5 \times 11$ and its aliquot parts 1, 2, 4, 5, 10, 11, 20, 22, 44, 55 and 110: total 284.

$284 = 2^2 \times 71$ and its aliquot parts are 1, 2, 4, 71 and 142, totalling 220.

According to Iamblichus, Pythagoras knew of this pair. However, Pythagoras may possibly not be the only ancient wise man to know of amicable numbers. Bible commentators point to Jacob's gift of 220 goats to Esau on their reunion – a friendly gift?

The brilliant Muslim mathematician, astronomer and physician Thabit ibn Qurra described in his *Book on the Determination of Amicable Numbers* Euclid's rule for perfect numbers, means of constructing abundant and deficient numbers, and the first rule for constructing amicable numbers, from which he deduced Pythagoras' pair, or perhaps more probably, the factors of 220 and 284 suggested the form of his rule: find a number, *n*, greater than 1, that makes these three expressions all prime:

$$a = 3 \times 2^n - 1 \quad b = 3 \times 2^{n-1} - 1 \quad c = 9 \times 2^{2n-1} - 1$$

Then the pair $2^n \times a \times b$ and $2^n \times c$ will be amicable.

The smaller of any Thabit pair is a tetrahedral number. 220 is the 10th tetrahedral. Lee and Madachy suggest that it may be significant that the first perfect number, 6, equals $1 \times 2 \times 3$; the smallest multiply perfect, 120, is $4 \times 5 \times 6$ and the sum of 220 and 284 is $504 = 7 \times 8 \times 9$. They comment that the Babylonians are known to have constructed tables of the products of 3 consecutive numbers, which are just 6 times the tetrahedral numbers.

There is an obvious similarity to Euclid's rule for even perfect numbers. However, Thabit's rule does not give all amicable pairs. Indeed, it is one of a number of similar patterns that generate amicable pairs. It is also very difficult to use, because it involves making 3 expressions prime simultaneously. Thabit ibn Qurra himself found no new pair. In fact his rule works for $n = 2$, 4 and 7, but for no other values below 20,000.

The 2nd pair, 17,296 and 18,416, was discovered by another Arab, Ibn al-Banna. It is Thabit's rule for $n = 4$. This pair was then rediscovered in 1636 by Fermat who also rediscovered Thabit's rule, as did Descartes who produced a 3rd pair, 9,363,584 and 9,437,056, two years later. This is the Thabit formula for $n = 7$.

Euler was the first mathematician successfully to explore amicable numbers and find many examples, more than 60. His methods are still the basis for present-day exploration.

More than 40,000 pairs of amicable numbers are now known, including all possible pairs in which the smaller number is less than a million.

Modern methods allow the generation of new amicable pairs from old. This often produces more than one 'daughter pair' per 'mother pair', which suggests that perhaps the number of amicable pairs is infinite.

Clearly the greater member of an amicable pair is deficient. Also, neither member of an even–even pair is divisible by 3.

In every case the numbers in a pair are either both even or both odd, though no reason is known why an even–odd pair should not exist.

Every pair also has a common factor. It is not known if a pair of coprime amicable numbers exists. If it does, then even in the most favourable case, in which their product is divisible by 15, that product itself must exceed 10^{67}. If they do, they will not of course be constructed on Thabit's pattern, or any similar pattern.

Most known amicable pairs have both numbers in the pair divisible by 3. However, this is not a general rule: this counter-example by te Riele may be the smallest such: $5 \times 7^2 \times 11^2 \times 13 \times 17 \times 19^3 \times 23 \times 37 \times 181$ multiplied by either $101 \times 8643 \times 1,947,938,229$ or by $365,147 \times 47,303,071,129$.

In 1968 Martin Gardner noticed that the sum of every even pair was divisible by 9 and naturally conjectured that this too was always so. It isn't, but counter-examples are rather rare; Elvin Lee gave the example 666,030,256, 696,630,544, originally discovered by Poulet.

Most amicable numbers have many different factors. Is it possible for a power of a prime, p^n, to be one of an amicable pair? If it is, then p^n is greater than 10^{1500} and n is greater than 1400.

A generalization of amicable pairs is amicable triplets, in which the proper divisors of any one number sum to the sum of the other 2. Beiler gives this example: $2^5 \times 3 \times 13 \times 293 \times 337$; $2^5 \times 3 \times 5 \times 13 \times 16,561$; $2^5 \times 3 \times 13 \times 99,371$.

[Lee and Madachy, JRM v5]

223

The largest number which cannot be represented with less than 37 5th powers.

232

232, 233 and 234 is the smallest triple of consecutive numbers each of which is the sum of 2 squares, and therefore the hypotenuse of a Pythagorean triangle:

$$232 = 6^2 + 14^2 \qquad 233 = 8^2 + 13^2 \qquad 234 = 3^2 + 15^2$$

It is not possible to have 4 such consecutive numbers.

239

$239 = 2 \times 4^3 + 4 \times 3^3 + 3 \times 1^3$

Together with 23, the only numbers that cannot be represented in fewer than 9 positive cubes. It also needs 19 4th powers to represent it.

If n is greater than 239, the largest prime factor of $n^2 + 1$ is at least 17. The number 239 fails, because $239^2 + 1 = 2 \times 13^4$. [Guy 5]

The only solution of the equation $x^2 + 1 = 2y^4$ is $x = 239$, $y = 13$.
[Lunggren quoted by Steiner and Tzanakis, Guy 153]

The 'approximation' to a Fermat equation, $x^4 + y^4 = z^4 + 1$, has 3 solutions with $x = 239$. The other numbers are $y = 104$, $z = 58,136$; $y = 143, z = 60,671$; $y = 208, z = 71,656$. [McLean, *Mathematical Spectrum* v18]

240

No number below 1,000,000 can have more than 240 divisors. 5 numbers have this many:

$$720,720 = 2^4 \times 3^2 \times 5 \times 7 \times 11 \times 13$$
$$831,600 = 2^4 \times 3^3 \times 5^2 \times 7 \times 11$$
$$942,480 = 2^4 \times 3^2 \times 5 \times 7 \times 11 \times 17$$
$$982,800 = 2^4 \times 3^3 \times 5^2 \times 7 \times 13$$
$$997,920 = 2^5 \times 3^4 \times 5 \times 7 \times 11$$

242

The numbers 242, 243, 244 and 245 each have 6 divisors.

The next such sequence of 4 consecutive numbers with equal numbers of divisors starts at 3655.

243

$243 = 3^5$, and is therefore 100,000 in base 3.

250

$250 = 5^3 + 5^3 = 13^2 + 9^2 = 15^2 + 5^2$ is the second sum of 2 cubes which is also the sum of 2 squares in more than one way. [Thayer]

251

The smallest number that is the sum of 3 different cubes in 2 ways: $251 = 1^3 + 5^3 + 5^3 = 2^3 + 3^3 + 6^3$.

256

$256 = 2^8$ or 100,000,000 in binary and 100 in hexadecimal.

$256 = 3^5 + 3^2 + 3 + 1$. Erdös has conjectured that no higher power of 2 is a sum of distinct powers of 3.

257

$257 = 4^4 + 1$ and is prime. The only known primes of the form $n^n + 1$ are when $n = 1, 2$ and 4. It has been shown that if there are other primes of this form, they must have more than 300,000 digits. [Madachy]

257 is therefore also of the form $n^2 + 1$. The sequence of primes of this form starts: 2 5 17 37 101 197 257 401 577 677 1297 . . . [Sloane 1506]

An example of a prime which is the average of its two neighbouring primes. The sequence of such primes starts: 5 53 157 173 211 257 263 . . . [Sloane 4011]

Fermat numbers

257 is the 3rd Fermat number, equal to $2^{2^3} + 1$.

Fermat in 1640, writing to Frenicle, stated that $2^n + 1$ is composite if

n is divisible by an odd number, and then asserted that every number $F_n = 2^{2^n} + 1$ is prime, although he could not prove this. He later sent the problem to Pascal, commenting, 'I wouldn't ask you to work at it if I had been successful.' Pascal did not take it up and it was Euler who first showed that Fermat was wrong.

The first 4 values, starting with F_0, are prime: $F_0 = 2^1 + 1 = 3$; $F_1 = 2^2 + 1 = 5$; $F_2 = 2^4 + 1 = 17$; $F_3 = 2^8 + 1 = 257$, and it is not difficult to show that $F_4 = 2^{16} + 1 = 65,537$ is prime, but the problem from then on is far more difficult, because the numbers increase in size so rapidly, more rapidly than any sequence that mathematicians had previously studied.

Fermat turned out to be mistaken, the only occasion on which he is known to have been wrong in his conjectures.

Euler, in 1732, showed that $F_5 = 2^{32} + 1 = 4,294,967,297 = 641 \times 6,700,417$. In 1747 he showed that any factor of a Fermat number F_n is of the form $k \times 2^{n+1} + 1$, which leads very quickly to the same factorization of F_5. He also found the same factor by using binary notation, one of the first uses of binary numbers in a mathematical proof. Over a century later, in 1880, F. Landry, who factorized many numbers of the forms $2^n + 1$ and $2^n - 1$, showed that $F_6 = 2^{64} + 1$ is the product of 2 primes: 274,177 and 67,280,421,310,721.

However, Pervushin had already discovered in 1877 that F_{12} is divisible by $7 \times 2^{14} + 1 = 114,689$. The Fermat numbers, like the Mersenne numbers, had already become an ideal testing ground for primality tests and methods of factorization.

To return to Fermat, it already appeared plausible that he was, unfortunately, totally wrong, that there are no prime Fermat numbers beyond F_4. Ideally, mathematicians sought for a complete factorization into primes, but often had to be satisfied, at least initially, with finding one factor, or proving that a particular F_n was composite, without actually producing any factor at all.

Thus, in 1909, Moorhead and Western proved that F_7 and F_8 are composite, without producing any factors. Such tests are easily performed today on computers using this criterion, which is similar to Lucas's test for the primality of Mersenne numbers: F_n is prime if and only if it divides $3^{1/x(F_n - 1)} + 1$.

The problem of F_7, which has 39 digits, illustrates very well the difference between using such a test and actually finding a factor. Not until 1970 did Morrison and Brillhart find its 2 prime factors: $F_7 = (2^9 \times 116,503,103,764,643 + 1)(2^9 \times 11,141,971,095,088,142,685 + 1)$.

In contrast, a factor of the giant F_{1945} is known, and more recently, in 1980, it was announced that $19 \times 2^{9450} + 1$ is a factor of F_{9448}. As Coxeter remarks, F_{1945} could never actually be written down because the number of digits far exceeds Eddington's estimate of the number of particles in the entire universe! How large then is F_{9448}? Yet it can be defined – it just has been – using only 5 symbols.

Fermat numbers are now known to be composite for all n from 5 to 21 inclusive and for many larger values of n, though only F_5, F_6, F_7, F_8, F_9 and F_{11} have been completely factored.

The smallest Fermat numbers whose status (prime or composite?) is undecided are F_{22}, F_{24} and F_{28}. [Ribenboim 73]

F_8 was conquered in 1981 when Brent and Pollard found the prime factor, 1,238,926,361,552,897, for which they suggested the mnemonic: 'I am now entirely persuaded to employ the method, a handy trick, on gigantic composite numbers.' The handy trick refers to their use of a Monte Carlo method, which as the name suggests uses a sophisticated version of throwing dice to discover the missing factor. How charming when chance is used to find a very definite number!

Fermat numbers have other properties, apart from being apparently almost all composite.

No F_n is triangular, except $F_0 = 3$, and no Fermat number is a square or a cube.

Gauss proved that a regular polygon with a prime number of sides can be constructed only if that number is a Fermat prime. Paucker gave the equations for constructing a regular 257-gon in 1822.

265
Subfactorial 6

267
$\sigma(267) = \sigma(295) = \sigma(323) = 360$, the smallest triplet in arithmetic progression.

276
Sociable chains and aliquot sequences
In a sociable chain the sum of the divisors of each number, excluding itself, leads to the next number and so on, eventually returning to the starting number. What happens if an arbitrary number is taken, and its sum of divisors calculated, and then the sum-of-divisors of the result, and so on?

Such a sequence is called an aliquot sequence. Some aliquot sequences may increase, on average, for ever. Some will enter a sociable chain and revolve for ever. Curiously, many aliquot sequences end up in Paganini's amicable pair 1184, 1210.

Catalan and then Dickson conjectured that all such sequences are bounded, though, according to Guy, heuristic arguments and experimental evidence suggest that some sequences, perhaps almost all those starting with an even number, go to infinity. Lenstra has proved that there are arbitrarily long monotonic increasing aliquot sequences.

te Riele has produced a sequence of this type that increases for more than the first 5000 terms.

276 is a test case for the conjecture. It is the smallest number whose final destination is unknown since D. N. Lehmer showed that 138, after rising to 179,931,895,322 after 117 steps, reached 1 after 177 steps. [Guy]

Lehmer and others have shown that 276 after 469 steps has produced the 45-digit number, 149,384,846,598,254,844,243,905,695,992,651, 412,919,855,640. More recently this calculation has been pushed to 487 steps.

What happens 'in the end'? No one knows.

281

The final repeated digit-products of consecutive integers 281, 282 and 283 are all 6: for example, $2 \times 8 \times 3 = 48$; $4 \times 8 = 32$; $3 \times 2 = 6$. There are no 4 consecutive integers with the same final digit-product under 6.24×10^9. [Lanska and Ashbacher, JRM v22 70]

284

With 220, the first pair of amicable numbers.

288

The 4th superfactorial: $1! \times 2! \times 3! \times 4!$. The sequence of superfactorials starts: 1 2 12 288 34,560 24,883,200 125,411,328,000 . . .

With 289, the second pair of consecutive powerful numbers: $288 = 2^5 3^2$ and $289 = 17^2$. The smallest such pair is 8, 9 and the next two pairs are 9800, 9801 and 332,928, 332,929.

The smallest multiple of 9 with entirely even digits. [A. J. Turner]

297

Kaprekar numbers

The 5th Kaprekar number. When an n-digit Kaprekar number is squared and the right-hand n digits are added to the left-hand n or $n - 1$ digits, the result is the original number: $297^2 = 88209$ and $88 + 209 = 297$.

The first few Kaprekar numbers are 1, 9, 45, 55, 99, 297, 703, 999, 2223, 2728, 4950, 5050, 7272, 7777 . . .

Note that $1 + 9 = 10$, $45 + 55 = 100$ and so on.

142,857 is Kaprekar. So is 1,111,111,111, the smallest Kaprekar number of 10 digits whose square is 12,345,671,900,987,654,321.

If a cyclic permutation of a Kaprekar number is squared and the 'halves' added, the result is a cyclic permutation of the original number. For example, 972 is a cyclic permutation of 297. $972^2 = 944,784$ and $784 + 944 = 1728$. The 'adding halves' process must now be completed by adding 1 to 728. The result, 729, is another, different, cyclic permutation of 297.

Similarly, 7272 is Kaprekar; its only distinct cyclic permutation is 2727: $2727^2 = 7,436,529$ and $743 + 6529 = 7272$.

297 is also a Kaprekar 'triple', because $297^3 = 026,198,073$ and $026 + 198 + 073$ also equals 297.

Kaprekar numbers are related to repunits. If the n-digit number X is Kaprekar, then $X^2 - X$ is a multiple of the n-digit repunit $(10^n - 1)/9$.

306
R. William Gosper, wishing to choose a number more or less at random as a test for a new method of calculating roots based on continued fractions, picked on 306 and calculated its 7th root to 2,800 digits. It starts, $2 \cdot 26518 \ldots$ [Gruenberger, 'Computer Recreations', *Scientific American*, April 1984]

313
The only 3-digit palindromic prime to be palindromic also in base 2. It equals 100111001_2. [M. E. Larsen]

319
319 cannot be represented as the sum of fewer than 19 4th powers.

325
$325 = 5 \times 5 \times 13$ is the smallest number to be the sum of 2 squares in 3 different ways: $1^2 + 18^2$, $6^2 + 17^2$ and $10^2 + 15^2$.

331
Given any number M, there is a power of 2, say 2^n, such that $M - 2^n$ or $M + 2^n$ has only prime factors greater than or equal to 331. [Cohen and Selfridge, MOC v29]

341
$341 = 11 \times 31$ is the smallest pseudoprime to base 2. That is, $2^{340} - 1$ is divisible by 341, although 341 is composite not prime.

The ancient Chinese believed that if n divides $2^{n-1} - 1$, then n is prime. So did Leibniz, but this is not so, as Pierre Sarrus first pointed out. Pseudoprimes are quite rare. There are 882,206,716 primes less than

20,000,000,000. In the same range Selfridge and Wagstaff calculate that there are only 19,865 pseudoprimes to base 2. [Pomerance, *Scientific American*, Dec. 1982]

353

353^4 is the smallest 4th power that is the sum of 4 other 4th powers, discovered by Norrie in 1911: $353^4 = 30^4 + 120^4 + 272^4 + 315^4$. The sequence of such numbers continues: 651 2487 2501 2829 ... [MOC v27 492]

360

The number of degrees in a full circle. The Zodiac was first divided into 360, no doubt by the division of each of the 12 signs into 30 equal parts.

The Greek astronomer Hipparchus first divided a general circle into 360 degrees.

Approximately the number of days in one year, divided roughly into 12 months of 30 days each.

365

The smallest number which is the sum of both 2 and 3 consecutive non-zero squares. $365 = 10^2 + 11^2 + 12^2 = 13^2 + 14^2$. The next is 35,645. Notice that all 5 squares are also consecutive, as in the equation $3^2 + 4^2 = 5^2$. The next such equation is: $21^2 + 22^2 + 23^2 + 24^2 = 25^2 + 26^2 + 27^2$.

365·24219 878

The calendar

The approximate number of days in a year, equal to 365 days 5 hours 48 minutes and 45·9747 seconds. This is the time taken for the earth to make one revolution of the sun. Every civilization has related it to the period of the moon's phases, for example the time between 2 new moons, which is approximately 29·530588 days, or 29 days 12 hours 44 minutes and 2·8 seconds.

Unfortunately, the relation cannot be a very simple one. It is coincidental that the length of the year in days is so close to the very round number, 360, which happens to be very close to 12 times the period of the moon. Such coincidences are helpful, but not enough, and immense ingenuity has been devoted to accounting for the differences.

In the Julian calendar the ordinary years have 365 days but every year whose number is divisible by 4 has an extra day, the 29th February, making a total of 366 days. The average Julian year has therefore 365·25 days and is one day out approximately every 128 years.

The Gregorian calendar, which is used today in most parts of the world, is a small but significant improvement on the Julian. All years divisible by 100 are ordinary years, not leap years, with the exception of years divisible by 400, which remain leap years. The Gregorian calendar contains one too many days every 3320 years, and so will not require adjustment until long after we are all dead.

The Julian and Gregorian calendars are based on the length of the year and therefore on the sun. Given any day of the year, we can tell fairly accurately the position of the sun in the sky, but not the position of the moon. The Muslim calendar in contrast gives the moon precedence. it has 12 months of alternately 30 and 29 days. In a leap year the last month has an extra day. The ordinary year has only 354 days and a leap year 355 days, so the start of the Muslim year moves steadily through the Gregorian year, and conversely.

The Jewish year is a combination of solar and lunar years. The basic year is a lunar year of 12 months that are alternately of 30 and 29 days, but when the error amounts to a full month, a 13th month is inserted into that year. This makes it the most complicated by far of all calendars. The complications that are introduced when the solar year and the lunar month are considered together are well illustrated by the manner in which the date of Easter, which depends on the position of the moon, jumps around in the Christian year. The great Karl Friedrich Gauss demonstrated his insight into numbers by constructing simple formulae for calculating the date of the Christian Easter festival, and also, which is even more difficult, the date of the Jewish festival of the Passover.

[Schocken, *The Calculated Confusion of Calendars*, Vantage Press, 1976]

370

Like 371 and 153, equal to the sum of the cubes of its digits.

371

371 equals the sum of the cubes of its digits.

399

399 needs 19 4th powers to represent it.

The smallest Lucas–Carmichael number n, such that if p divides n, then $p + 1$ divides $n + 1$. The sequence of such numbers continues: 935 2015 2915 . . . [Sloane 5450]

400

$400 = 20^2 = 1 + 7 + 7^2 + 7^3$

In other words, the sum of the divisors of 7^3 is a square. The sum of the divisors of 400 is also a square: $961 = 31^2$.

400 is also the product of all the proper divisors of 20.

407

$407 = 4^3 + 0^3 + 7^3$

461

The number of primes of the forms $4n + 1$ and $4n - 1$, up to and including 461, are equal. They are equal at 2, 5, 17, 41, 461, 26833, 26849, 26863, 26881, 26893, 26921, 616769 . . . [Sloane 1507]

462

The second smallest number whose square ends in the digits 444: $462^2 = 213,444$.

466

The largest number which cannot be represented with less than 32 5th powers.

481

481^2 is the smallest square which is the sum of 3 biquadrates: $12^4 + 15^4 + 20^4$.

484

$484 = 22^2$

It is the palindromic square of a palindromic square root.

487

One of only 3 primes, less than 2^{32} such that p^2 divides $10p - 10$. The others are 3 and 56,598,313. [Montgomery, MOC v61 361]

495

Take any 3-digit number whose digits are not all the same and which is not a palindrome. Arrange its digits in ascending and descending order and subtract. Repeat. This is called Kaprekar's process. All 3-digit numbers eventually end up with 495, and stick there, since $954 - 459 = 495$.

496

The 3rd perfect number. $496 = 16 \times 31 = 2^4(2^5 - 1)$ is equal to the sum of all its proper divisors, $1 + 2 + 4 + 8 + 16 + 31 + 62 + 124 + 248 = 496$.

Thomas Greenwood noticed that 1 more than an even or 2 less than an odd triangular number whose index is prime is often a prime number. $T_{31} = 496$, the 31st triangular number, is the first counter-example. 31 is prime, but 497 is divisible by 7.

499

$499 = 497 + 2$ and $497 \times 2 = 994$, its reversal.

500

Denoted by the letter D in Roman numerals.

504

504 is equal to both 12×42 and 21×24.

There are thirteen such 2-digit pairs, the largest being $36 \times 84 = 63 \times 48 = 3024$.

510

$7^{510} = 1 \cdot 00000\,09377\,76536 \ldots \times 10^{431}$, the closest known approximation of a power of 7 by a power of 10.

512

$512 = 2^9$

It is therefore, 1,000,000,000 in binary and 1,000 in octal.

512·73

Numerology

512·73 is the Dewey Decimal classification, under the general class '510 mathematics' for 'number theory: analytic'. When Martin Gardner wrote *The Numerology of Dr Matrix*, the Dewey classification for 'number theory' was, as he pointed out, 512·81, whose two halves are respectively 2^9 and 9^2.

No doubt because this trivial piece of numerology has been found out, the authorities have since changed 'number theory' to 512·7 and given this new number, 512·73, to analytic number theory, whose first classification is, significantly, transcendental numbers. I shall now illuminate the profound significance of 512·73 for the benefit of the uninitiated.

First, I subtract it from 666, the Number of the Beast in the Book of

Revelation: $666 - 512 \cdot 73 = 153 \cdot 27$. Behold! The same digits appear, but rearranged, symbolizing the effect of removing evil from the world. The first number is now 153, the number of fishes hauled from the sea by Peter, which was so eloquently interpreted by St Augustine. The second number is now the sacred number 3, raised to its own power. 153 is also associated with the sacred number 3. Not only is its sum of digits equal to 9, which is 3 times itself, but it is the sum of the 3rd power of its own digits. The significance of 3 appears in the Dewey Decimal System. Divide the Number of the Beast by 3, and you obtain 222, the classification of the Old Testament. Add 3, and you obtain 225, the New Testament. Add 3 again, and you obtain 228, which is the Book of Revelation.

And so on, and on, and on, and on, and on . . .

I trust this illustrates how an hour's worth of jiggery-pokery with a selection of numbers (choose the ones you want, ignore the rest) will produce out of that hat any number you desire . . .

518
$518!! + 1 = 518 \times 516 \times 514 \times 512 \times 510 \times 508 \ldots 6 \times 4 \times 2 + 1$ is prime. 518!!, as defined here, is double-factorial 518. The double factorial of $517 = 517 \times 515 \times 513 \ldots 5 \times 3 \times 1$

$n!! + 1$ is prime for no other values of n up to 2596, apart from $n = 1$ and $n = 2$. [*Mathematical Spectrum* v26]

521
A prime Lucas number: the sequence of these starts: 3 7 11 29 47 199 521 2207 3571 9349 3,010,349 . . . [Sloane 2627]

523
$\phi(523) = 522$, $\phi(524) = 260$, $\phi(525) = 240$, the first sequence of three successive strictly decreasing values of $\phi(n)$. 525, 526, 527 form a sequence of strictly increasing values.

559
The largest number less than 4100 which cannot be represented as the sum of less than 19 4th powers. The others are 79, 159, 239, 319, 399 and 379.

561
$561 = 3 \times 11 \times 17$ is the smallest Carmichael number, otherwise called an absolute pseudoprime, meaning that it is a pseudoprime to any base at all. In other words, $a^{560} - 1$ is divisible by 561, whatever the value of a, provided 561 and a are coprime.

R. D. Carmichael proved in 1912 that every Carmichael number is the product of at least 3 odd primes. It is now known that n is a Carmichael number if and only if it is the product of at least 3 different odd primes, $p_1, p_2, p_3 \ldots$, and for every one of these factors $n - 1$ is divisible by $p - 1$. It is widely believed, but not proved, that there are an infinite number of Carmichael numbers, though they are rare.

The sequence of Carmichael numbers continues: 1105 1729 2465 2821 6601 8911 10,585 . . .

563

According to Wilson's theorem, $(p - 1)! + 1$ is divisible by p if and only if p is prime. Very occasionally, it is divisible by p^2. The only such values below 4,000,000 are 5, 13 and 563. This was found by Goldberg in 1953, a very early example of the use of computers to tackle a problem in number theory.

567

$567^2 = 321,489$

This equation uses each of the digits 1 to 9, once each.

The only other number with this property is 854.

587

The start of a sequence of 11 primes, formed by trebling each number in turn and adding 16. [JRM v13]

587 1777 5347 16,057 48,187 144,577 433,747 1,301,257 3,903,787 11,711,377 35,134,147

593

Wilhelm Fliess, a friend and correspondent of Sigmund Freud, believed that just about any phenomenon in the world could be explained by combinations of the numbers 23 and 28. If he had been a better mathematician, he would have realized that all but a finite set of numbers can be represented in the form $23n + 28m$, n and m both positive. $593 = 23 \times 28 - 23 - 28$ happens to be the largest number that cannot be so represented.

There is nothing special about the choice of 23 and 28. Any two numbers that have no common factor may be chosen.

625

$625 = 5^4$

Because 625^2 ends in the same digits, 390,625, any power of 625 ends in the same digits.

$5^4 = 2^4 + 2^4 + 3^4 + 4^4 + 4^4$ is the smallest 4th power to be the sum of 5 other 4th powers.

641

Euler found the first counter-example to Fermat's conjecture that $2^{2^n} + 1$ is always prime, when he discovered in 1742 that $2^{2^5} + 1$ is divisible by 641.

All factors of $2^{2^n} + 1$ are of the form $k \times 2^{n+1} + 1$. In this case, $641 = 10 \times 2^6 + 1$.

645

The 3rd smallest pseudoprime to base 2: $2^{644} + 1$ is divisible by 645 although $645 = 3 \times 5 \times 43$ is composite.

651

$651^4 = 240^4 + 340^4 + 430^4 + 599^4$ is the second smallest solution, following 353^4, for a 4th power as the sum of 4 other 4th powers.

651 has the unusual property that 651×156 is equal to another product of the same pattern, 372×273. [Iyangar, *Scripta Mathematica*, 1939]

661

The start of a record-breaking gap in the sequence of twin primes. 659 and 661 are twin primes; the next pair is 809, 811. [Sloane 641]

666

The 36th triangular number ($666 = \frac{1}{2} \times 36 \times 37$) and the Number of the Beast in the Book of Revelation: 'Here is wisdom. Let him that hath understanding count the number of the beast; for it is the number of a man, and his number is six hundred, three score and six.'

A number beloved of occultists, who throughout the ages have used gematria to find the Number of the Beast in the names of their enemies, political or theological. The fact that some ancient authorities give the number as 616 has not deterred them. With a little ingenuity, both numbers can be found instead of just one.

Peter Bungus made Luther equal to 666, by using the old system, which counts A–I as 1–9, K–S as 10–90, and T–Z as 100–500. Bungus read Luther's name as Martin Luthera, half German and half Latin, a typical bit of skulduggery, but Bungus was an expert. He wrote a dictionary of numerological symbolism.

666 in Roman numerals is D C L X V I, which uses each letter under M (1000) once, which has led to the suggestion that this is the origin of 666.

It could merely be a way of expressing some large, or vague, number. $-2 \sin(666)$ is a good approximation to the Golden Ratio, ϕ. [Wang, JRM v26 203]

$\phi(666) = 216 = 6 \times 6 \times 6$.

$666 = 2^2 + 3^2 + 5^2 + 7^2 + 11^2 + 13^2 + 17^2$, the sum of the squares of the first seven primes.

672

The 2nd triperfect number, after 120: $672 = 2^5 \times 3 \times 7$ and the sum of its divisors is $3 \times 672 = 2016$.

675

With 676, the only known pair of consecutive powerful numbers in which the smaller is odd. $675 = 3^3 \times 5^2$ and $676 = 2^2 \times 13^2$. [AMM v77 850]

676

The smallest palindromic square whose square root is not palindromic: $676 = 26^2$.

679

The smallest number with multiplicative persistence equal to 5.

The product of its digits is 378, the product of whose digits is 168, which generates 48, which generates 32, which generates 6, a total of 5 steps.

680

680 is the smallest tetrahedral number to be the sum of two tetrahedral numbers: $680 = 120 + 560$.

714

On April 8, 1974, in Atlanta, Georgia, Henry Aaron hit his 715th major league homerun, thus eclipsing the previous mark of 714 long held by Babe Ruth. This event received so much advance publicity that the numbers 714 and 715 were on millions of lips. Questions like, 'When do you think he'll get 715?' were perfectly understood, even with no mention made of Aaron, Ruth or homerun. In all the hubbub it appears certain interesting properties of 714 and 715 were overlooked . . .

wrote C. Nelson, D. E. Penney and C. Pomerance in *The Journal of Recreational Mathematics*, 1974. The authors note some very unusual properties indeed. First, $714 \times 715 = 510{,}510 = 2 \times 3 \times 5 \times 7 \times 11 \times 13 \times 17$, which is primorial 17, the product of all the primes up to and including 17. They discovered on computer that only primorial 1, 2,

3, 5 and 7 can be represented as the product of consecutive numbers, up to primorial 3049.

They further notice that σ(714), which is defined as the sum of the divisors of 714 including itself, is a perfect cube, and that the ratio σ(714)/φ(714) is a perfect square. Finally they notice that 714 + 715 = 1429 which has the property that 6 arrangements of its digits are prime numbers.

720

$720 = 6!$ and is also the product of consecutive integers in 2 ways: $720 = 10 \times 9 \times 8 = 6 \times 5 \times 4 \times 3 \times 2$.

This is equivalent to saying that $10! = 7! \times 6!$, the only known example of a factorial being the product of two other factorials.

$6! = 24^2 + 12^2$. No larger factorial is the sum of two squares.

$720!$ is divisible by 720^{178}.

728

This number and 729 are consecutive Smith numbers. [Dudley, MM v67]

729

9^3 and the 2nd smallest cube to be the sum of 3 cubes: $9^3 = 1^3 + 6^3 + 8^3$. This makes 7^3 the smallest solution to the approximate Fermat equation, $x^3 = y^3 + z^3 + 1$. The next such solution is $103^3 = 64^3 + 94^3 + 1$.

Since $6^3 = 3^3 + 4^3 + 5^3$, 9^3 is also the sum of 5 cubes.

$729 = 3^6$ and therefore is 1,000,000 in base 3.

729 is another mysterious number in Plato's *Republic*:

... if one were to express the extent of the interval between the king and the tyrant in respect of true pleasure he will find on completion of the multiplication that he lives 729 times as happily and that the tyrant's life is more painful by the same distance.

729 was of great significance to the Pythagoreans, being 27^2. Plato combined the two sequences of powers of 2 and 3 as far as the cubes to form the sequence 1 2 3 4 8 9 27. Here 27 is the sum of all the preceding members.

C. A. Browne interprets the number in terms of a magic square 27 by 27, whose central cell is occupied by 365, the number of days in the year ($729 = 364 + 365$).

1/729 has a decimal period of 81 digits, which can be arranged in groups of 9 digits, reading across each row, in this pattern [Thébault, *Scripta Mathematica*, v19]:

```
001 371 742
112 482 853
223 593 964
334 705 075
445 816 186
556 927 297
668 638 408
779 149 519
890 260 631
```

735

735 is divisible by the product of its digits, which is larger than that of any smaller number with the same property. [Sloane 482]

767

$$\binom{767}{1} + \binom{767}{2} + \binom{767}{3} + \binom{767}{4}$$ is a perfect square, 8672^2. There is

no larger solution. The smaller solutions are 7, 15 and 74. [Guy 147]

780

780 and 990 are the 2nd smallest pair of triangular numbers whose sum and difference (1770 and 210) are also triangular.

   The side of the smallest square which can be inscribed in two separate Pythagorean triangles, to rest on the hypotenuse. The triangles have sides 1924, 1443, 2405 and 1145, 2748, 2977. [Korbin]

788

The numbers 788, 789, 790, 791, 792, 793 are divisible by 2, 3, 5, 7, 11 and 13 respectively.

818

818 to 831 is the largest gap between 2 semi-primes less than 1000.

836

Almost all palindromic squares seem to have an odd number of digits. 836 is the first with an even number: $836^2 = 698,896$. It is also the largest number below 1000 whose square is palindromic. The next such square is $798,644^2 = 637,832,238,736$. [Ondrejka, JRM v20 71]

840

$840 = 2^3 \times 3 \times 5 \times 7$

It is the number below 1000 with the largest number of divisors: $2^5 = 32$.

873

$873 = 1! + 2! + 3! + 4! + 5! + 6!$

880

There are exactly 880 magic squares of order 4, provided that all rotations and reflections of the same square are counted as one.

888

The smallest multiple of 24 whose digits sum to 24. [Sloane 489]

945

The first odd abundant number, discovered by Bachet. It is also semi-perfect.

$945 = 3^3 \times 5 \times 7$ and its divisors sum to 975.

Odd abundant numbers are quite rare. There are only 23 of them below 10,000.

981

The only known example of 5 triplets of numbers such that the sums of each triplet are equal and their products also are equal, is: 6, 480, 495; 11, 160, 810; 12, 144, 825; 20, 81, 880; 33, 48, 900.

The sum of each triplet is 981, and their common product 1,425,600.

999

The minimum sum of pandigital 3-digit primes, $149 + 263 + 587 = 999$.

The smallest multiple of 27 whose digits sum to 27.

$999^2 = 998,001$ and $998 + 001 = 999$, so 999, like all numbers whose digits are all 9s, is Kaprekar.

In fact, any multiple at all of 999 can be separated into groups of 3 digits from the unit position, which when added will total 999.

The same principle applies to multiples of 9 99 9999 and so on.

$999 = 27 \times 37$ and so $1/27 = 0 \cdot 037037 \ldots$ and $1/37 = 0 \cdot 027027 \ldots$

1000

$1000 = 10^3$ in any base at all.

1001

$1001 = 7 \times 11 \times 13$

This is the basis for a test of divisibility that will test for all 3 divisors simultaneously. Mark off the number to be tested in groups of 3 digits from the unit position. Large numbers are more often than not already written in this manner, for example, 68,925,857. Add the 1st, 3rd, 5th groups and take away the total of the 2nd, 4th . . . groups. The number will be divisible by 7, 11 or 13, if the result is divisible by 7, 11 or 13 respectively: in this example, $68 + 857 - 925 = 0$.

1001 is the 4th palindromic pentagonal number. The sequence of such numbers runs: 1 5 22 1001 2882 15,251 720,027 . . . [Sloane 3924]

1009

The smallest prime that can be expressed in the form $x^2 + ny^2$ for all values of n from 1 to 10. The next such numbers are 1129 and 1201. [Ashbacher, JRM v24 202]

1024

$1024 = 2^{10}$ and therefore the smallest number with 10 prime factors.

Although kilo- in the metric system usually means a thousand, as in kilogramme, 1K of memory in a computer means 1024. It is a neat coincidence that 2^{10} is so close to 10^3.

1056

Equal to $32 \times 33 = 1 \times 6 \times 11 \times 16$, one of only two solutions of $x(x + 1) = y(y + 5)(y + 10)(y + 15)$. The other is $207 \times 208 = 8 \times 13 \times 18 \times 23 = 43,056$. [*Acta Arithmetica* v7 194]

1089

$1089 \times 9 = 9801$

The same property is true of 10,989, 109,989 and so on.

$1/1089 = 0 \cdot 0009182736455463728191\ 00091 \ldots$ [N. Goddwin]

If a 3-digit number is reversed and the result subtracted, and that answer added to its reversal, the answer is always 1089: $623 - 326 = 297$ and $297 + 792 = 1089$.

Note the middle digit, 9, and the fact that $1089 = 999 + 90$.

The only other number of 4 or fewer digits whose reversal is a multiple of itself is $2178 = 2 \times 1089$. This number and 1089 were cited by G. H. Hardy as examples of non-serious mathematics.

$1089 = 33^2 = 65^2 - 56^2$.

This is the only 2-digit example of this pattern.

1091

The polynomial $x^6 + 1091$ is composite for all values of x from 1 to 3905. [Shanks, Ribenboim 330]

1093

$2^{1092} - 1$ is divisible by 1093^2.

Only one other number is known below 4×10^{12} with this property, 3511.

In 1909 Wieferich created a sensation by proving that if Fermat's equation, $x^p + y^p = z^p$, has a solution in which p is an odd prime that does not divide any of x, y or z, then $2^{p-1} - 1$ is divisible by p^2. As facts about Fermat's Last Theorem go, this is remarkably simple. That 1093 and 3511 are the only solutions below 4×10^{12} means that only these two cases of Fermat's theorem need to be considered, below that limit, if p does not divide xyz.

1105

$1105 = 5 \times 13 \times 17$ is the product of the first 3 primes of the form $4n + 1$.

It is the second Carmichael number. It can also be expressed as the sum of two squares in more ways than any smaller number. [Guy 141]

1111

$1111 = 56^2 - 45^2$ following $11 = 6^2 - 5^2$. The pattern continues, $556^2 - 445^2 = 111,111$ and so on.

Similarly, $7^2 - 4^2 = 33$, $67^2 - 34^2 = 3333$ and so on and $8^2 - 3^2 = 55$, $78^2 - 23^2 = 5555$ and so on . . .

1127

Thébault gives $1127^2 = 01,270,129$ as an example of a square that consists of 2 consecutive odd or even numbers juxtaposed.

He quotes the 'matching' number $8874^2 = 78,747,876$, noting that $1127 + 8874 = 10,001$, and other pairs with the same property. [Thébault, *Scripta Mathematica* v13]

1141

$1141^6 = 74^6 + 234^6 + 402^6 + 474^6 + 702^6 + 894^6 + 1077^6$

This is the smallest known solution for a 6th power as the sum of 7 other 6th powers.

1167

The largest of 256 numbers which *cannot* be expressed as the sum of at most 5 composite numbers. [Guy 136]

1184

With 1210, the 2nd smallest pair of amicable numbers, discovered in 1866 by Nicolò Paganini when he was a 16-year-old schoolboy, having been previously missed by Descartes, Fermat, Euler and many others.

1210

With 1184, Paganini's pair of amicable numbers.

1225

$1225 = 35^2 = \frac{1}{2} \times 49 \times 50$

It is the second number to be simultaneously square and triangular. The next two are 204^2 and 1189^2.

$35^2 = 1^3 + 3^3 + 5^3 + 7^3 + 9^3$. The next such sum is $1^3 + \ldots 29^3$.

1233

$1233 = 12^2 + 33^2$

B. S. Rao finds such numbers by expressing a number of the form $n^2 + 1$ as the sum of 2 squares in another way.

Another example is $8833 = 88^2 + 33^2$. [MM v57]

1309

With 1310 and 1311, the first triple of consecutive integers, each the product of 3 distinct primes. The next two such triples are 1885, 1886, 1887 and 2013, 2014, 2015.

1331

$1331 = 11^3 = 19^2 + 21^2 + 23^2$. [Stajsczak]

1375

1375, 1376 and 1377 is the smallest triple of consecutive integers, each of which is divisible by a cube, apart from 1. [*Eureka*, 1982]

The smallest such quadruplet is 22,624 to 22,627. [Vandemergel]

1444

$1444 = 38^2$

It is the smallest square ending in the repeated digits ... 444. The next is $462^2 = 213,444$. The general formula is $(500n \pm 38)^2 = \ldots$ 444. [Beiler]

1444 is also the 5th square, not counting multiples of 100, whose digits form 2 other squares juxtaposed, $1444 = 144:4$. The others are 49, 169, 361 and 1225.

1477

1477! + 1 is prime. $n! + 1$ is also prime for n = 1, 2, 3, 11, 27, 37, 41, 73, 77, 116, 154, 320, 340, 399, 427, 872, 1477 . . . [Sloane 908]

1540

One of only 5 numbers that are simultaneously triangular and tetrahedral.

It is the 55th triangular number and the 20th tetrahedral.

1549

1549 is the only odd number below 10,000 that is not the sum of a prime and a power.

1634

$1634 = 1^4 + 6^4 + 3^4 + 4^4$. The other 4-digit numbers with this property are 8208 and 9474.

1681

$1681 = 41^2$

The only 4-digit square whose 2-digit 'halves' are also squares, apart from the obvious set, 1600, 2500 . . . 8100.

1728

$1728 = 12^3$ and therefore equal to 1000 in the duodecimal system, and the number of cubic inches in a cubic foot.

1729

Among the most famous of all numbers, due to an incident described by G. H. Hardy. Ramanujan, Hardy writes,

could remember the idiosyncrasies of numbers in an almost uncanny way. It was Littlewood who said that every positive integer was one of Ramanujan's personal friends. I remember going to see him once when he was lying ill in Putney. I had ridden in a taxi-cab No. 1729, and remarked that the number seemed to me a rather dull one, and that I hoped it was not an unfavourable omen. 'No,' he reflected, 'it is a very interesting number; it is the smallest number expressible as the sum of two cubes in two different ways.' [Hardy, *Ramanujan*, CUP, 1940]

$1729 = 12^3 + 1^3 = 10^3 + 9^3$

Hardy then asked Ramanujan whether he knew the answer to the same problem for 4th powers, Ramanujan thought for a moment, and replied that he did not, but it must be very large. [*See 635,318,657.*]

This property of 1729 was found by Frenicle, a brilliant calculator, who in reply to a challenge from Euler gave five solutions: $9^3 + 10^3 = $

$12^3 + 1^3$; $9^3 + 15^3 = 2^3 + 16^3$; $15^3 + 33^3 = 2^3 + 34^3$; $16^3 + 33^3 = 9^3 + 34^3$ and $19^3 + 24^3 = 10^3 + 27^3$.

1729 is also Harshad, that is, it is divisible by the sum of its own digits: $1729 = 19 \times 91$.

The smallest number that is a pseudoprime simultaneously to bases 2, 3 and 5. [Pomerance, Selfridge and Wagstaff, MOC v35 1003]

1729 is the largest number which factorizes into the sum of its digits, 19, and its reversal. There is one other 4-digit number with this property, $1458 = 18 \times 81$. [Trigg, JRM v9 44]

It is also the 3rd Carmichael number.

1760

1 statute mile = 1760 yards = 320 rods, poles or perches = 8 furlongs.

1782

1782 is equal to 3 times the sum of all the 2-digit numbers that can be made from its digits, 1, 7, 8 and 2.

1854

1854 is subfactorial 7, or !7.

1880

1880, 1881, 1882 is the first triple of consecutive happy numbers.

1980

$1980 - 0891 = 1089$. This is one of only 6 patterns in which subtracting a 4-digit number from its reversal leaves the digits rearranged.

The others are $7641 - 1467 = 6174$; $5823 - 3285 = 2538$; $3870 - 0783 = 3087$; $2961 - 1692 = 1269$ and $9108 - 8019 = 1089$.

The smaller of a triple of numbers, $1980 = 2^2 \times 3^2 \times 5$, $2016 = 2^5 \times 3^2 \times 7$, and $2556 = 2^2 \times 3^2 \times 71$, whose sums of divisors are equal to the sum of the three numbers. [Guy 59]

2025

$2025 = 45^2$ and $20 + 25 = 45$, which is therefore a Kaprekar number.

When each of its digits is increased by 1, 2025 becomes 3136, which is also a square, 56^2. A matching pair of 2-digit squares are 25 and 36.

2047

$2047 = 2^{11} - 1$, and is therefore the 11th Mersenne number. It is the first Mersenne number with a prime exponent, which is composite: $2047 = 23 \times 89$.

2178

$2178 \times 4 = 8712$, its own reversal. The same pattern works for 21,978, 219,978, and so on.

2178 is a 4th-order digital invariant, switching to 6514 and back: $2^4 + 1^4 + 7^4 + 8^4 = 6514$ and $6^4 + 5^4 + 1^4 + 4^4 = 2178$.

2185

The start of a sequence of 17 consecutive integers, each of which has a common factor, greater than 1, with the product of the remaining 16. [Guy 83]

2187

$2187 = 3^7$ or 10,000,000 in base 3.

2240

The number of pounds in an English ton. In America, the number of pounds in a long ton, the short ton being of 2000 lbs.

1 ton = 2240 lbs = 160 stone = 80 quarters = 20 hundredweight

2310

Primorial 11 = $2 \times 3 \times 5 \times 7 \times 11$ and therefore the smallest number with 5 different prime factors.

2333

$2333^2 = 5,442,889$, following the pattern $3^2 = 9$, $23^2 = 529$, $233^2 = 54,289$ and so on.

2401

$2401 = 7^4 = 227^4 + 157^4 - 239^4$. This is the smallest 4th power which can be represented in this form, with all numbers under 10^6. [Zajta, MOC v41 635]

2465

$2465 = 5 \times 17 \times 29$, and 4, 16 and 28 each divide 2464, so it is a Carmichael number, in fact the 4th.

2520

$2520 = 2^3 \times 3^2 \times 5 \times 7$ is the lowest common multiple of the digits 1 to 9, and the sum of 4 of its own divisors in 6 ways, the maximum possible.

The six combinations of factors are: 1260, 630, 504, 126; 1260, 630, 420, 210; 1260, 840, 360, 60; 1260, 840, 315, 105; 1260, 840, 280, 140; 1260, 840, 252, 168.

2520 is the smallest such number. The problem is equivalent to expressing unity as the sum of 4 reciprocals.

The largest of 6 numbers which share the property of each having more divisors than any lesser number, and also of dividing any larger number with the same property. The other 5 are 1, 2, 6, 12 and 60. [Ratering, MM v64 343]

$2520 = 120 \times 021 = 210 \times 012$. The smallest number which can be written as the product of a number and its reversal in 2 ways. [S. S. Gupta]

2538

Start as with the Kaprekar process (*see* 6174) but then switch the first two and last two digits, so that 4909 produces the two numbers 9904 and 4099. Subtract and repeat. The eventual result is always 2538. [Trigg, JRM v22 34]

2584

The 18th Fibonacci number, $F_{18} = 7^3 + 8^3 + 9^3 + 10^3$. [Stajsczak]

2592

$2592 = 2^5 \times 9^2$, the only pattern of its kind. [Dudeney]

2601

$2601 = 51^2$. This square can be represented as the sum of 3 squares in 10 distinct ways. The largest smallest-square solution is $24^2 + 27^2 + 36^2$. The smallest largest-square solutions are $34^2 + 31^2 + 22^2$ and $34^2 + 34^2 + 17^2$. [JRM v22 75]

2615

$2615 \times 11 = 28,765$ and $5162 \times 11 = 56,782$, its reversal.

The choice of 2615 for this property is highly arbitrary, because this pattern works whenever the adjacent digits of a number do not sum to more than 9, for any pair. So it works for 2,363,511,509 but not for 45,173, for example.

2620

The 1st member of the 3rd pair of amicable numbers. Its partner is 2924.

3000

The smallest number requiring more than 12 letters to write in the English language. It uses 13.

3003

3003 is the smallest number to appear 8 times in Pascal's triangle.

There is no other number appearing so often, less than 2^{23}. [Singmaster, AMM v78]

3334

$3334^2 = 11,115,556$, following the pattern $4^2 = 16$, $34^2 = 1156$, $334^2 = 111,556$ and so on.

When cubed, $3334^3 = 037059263704$ and the sum of the 3 4-digit numbers is $0370 + 5926 + 3704 = 10,000$. [JRM v14]

3367

This number can be multiplied by a 2-digit multiplier xy by dividing the number $xyxyxy$ by 3.

This trick works because $3367 = 10101/3$.

3435

$3435 = 3^3 + 4^4 + 3^3 + 5^5$. This is the only known number with this property, apart from $1^1 = 1$. [Madachy]

3511

One of only two known numbers that make $2^{p-1} - 1$ divisible by p^2.

The other is 1093.

3600

$3600 = 60^2$

The number of seconds in an hour, or seconds in a degree, or minutes in 60 degrees.

4096

$4096 = 2^{12} = 8^4 = 16^3$

It is therefore equal to $1,000,000,000,000$ in binary; $10,000$ in octal; and 1000 in hexadecimal.

4104

The smallest integer that can be represented as the sum of 2 cubes in 3 ways: $4104 = 16^3 + 2^3 = 15^3 + 9^3 = (-12)^3 + 18^3$.

4150

$4150 = 4^5 + 1^5 + 5^5 + 0^5$. The smallest number which is the sum of the 5th powers of its digits. [Lehning]

4181

The 19th Fibonacci number, but although 19 is prime, this is not: $4181 = 37 \times 113$. This is the first composite Fibonacci number with a prime index.

4356

4356 multiplied by $1\frac{1}{2}$ is 6534, its reversal. Note that $4356 = 1089 \times 4$.

4523

The smallest number which is non-palindromic in binary, but whose square in binary is palindromic:

$$1000110101011^2 = 1001110000010100000111001$$

4713

The index of the 3rd prime Cullen number: $4713 \times 2^{4713} + 1$ is prime. The 6 known prime Cullen numbers have indices, 1, 141, 4713, 5795, 6611, 18,496. [Sloane 5401]

4840

$4840 = 22 \times 220$ is the number of square yards in an acre.

4900

The only square pyramidal number that is also a square.
$4900 = 70^2 = 1^2 + 2^2 + 3^2 + 4^2 + \ldots + 24^2$.

4913

$4913 = 17^3$ and also the sum of its digits is 17. Other cubes with this property are $512 = 8^3$, $5832 = 18^3$, $17,576 = 26^3$, and $19,683 = 27^3$.

5020

The 1st member of the 4th amicable pair. Its friend is 5564.

5040

Factorial 7. $5040 = 7! = 1 \times 2 \times 3 \times 4 \times 5 \times 6 \times 7$. Also $5040 = 7 \times 8 \times 9 \times 10$, making it the product of consecutive integers in 2 ways.

In bell ringing, a complete sequence of Stedman Triples contains $7! = 5040$ changes, and takes 3 or 4 hours to ring.

Plato, in the *Laws*, suggested that a suitable number of men for an ideal city would be that number which contained the most numerous and most consecutive subdivisions. He decides on 5040, indicating that this number has 59 divisors (apart from itself) and can be divided for purposes of war 'and in peace for all purposes connected with contributions and distributions' by any number from 1 to 10.

Moreover, by merely subtracting two hearths from the total, it is then divisible exactly by 11 also.

5186

$\phi(5186) = \phi(5187) = \phi(5188) = 2^5 \times 3^4$

This is the first such triple of successive integers with the same ϕ values.

5282

The number of ways of placing 8 non-attacking rooks on a standard chessboard. [Sloane 1761]

5777

The conjecture that every odd number can be represented in the form $p + 2a^2$, where p is unity or a prime, is false, but there are only two counterexamples below 6×10^5.

5777 is one, and 5993 is the other. [Ashbacher, JRM v22 244]

5778

The only triangular Lucas number, apart from 1 and 3. [Ming, FQ v27]

5906

The smallest integer which is the sum of 2 rational 4th powers, but not the sum of 2 integer 4th powers. [Guy 145]

5913

$5913 = 1! + 2! + 3! + \ldots + 6! + 7!$

6174

Kaprekar's process

6174 is Kaprekar's constant, the result of Kaprekar's process applied to any 4-digit number, apart from the exceptional numbers whose digits are all equal.

Take any other 4-digit number, and arrange the digits in ascending and descending order, so that, for example, 4527 leads to 2457 and 7542. Subtract, and repeat. The eventual result is the number 6174:

$$7542 - 2457 = 5085$$
$$8550 - 0558 = 7992$$
$$9972 - 2799 = 7173$$

$$7731 - 1377 = 6354$$
$$6543 - 3456 = 3087$$
$$8730 - 0378 = 8352$$
$$8532 - 2358 = 6174$$
$$7641 - 1467 = 6174 \text{ and the calculation repeats.}$$

6174 is also a Harshad number, because it is divisible by the sum of its digits.

6578
$$6578 = 1^4 + 2^4 + 9^4 = 3^4 + 7^4 + 8^4$$

This is the smallest representation of a number as the sum of 3 4th powers in 2 ways.

6666
$6666^2 = 44435556$ and the two halves 4443 and 5556 sum to 9999.

The pattern is the same for any string of 6s. Compare $3333^2 = 11108889$ and $1110 + 8889 = 9999$, and $7777^2 = 60481729$ where $6048 + 1729 = 7777$, making 7777 Kaprekar.

More generally, if a number is multiplied by a number whose digits are all the same, for example, let 894 be multiplied by 22,222, then in this case the right-hand 5 digits, added to the left-hand portion, form another number with equal digits: $894 \times 22,222 = 19866468$ and $198 + 66,468 = 66666$.

6667
$6667^2 = 44,448,889$ and $44,448,889 \times 3 = 133,346,667$, which ends in the same 4 digits, 6667. Hence 6667 is called tri-automorphic.

For any given number of digits, there are 3 tri-automorphic numbers. The others for 4 digits are 9792 and 6875.

The 3 10-digit tri-automorphic numbers are 6,666,666,667, 7,262,369,792 and 9,404,296,875. [Hunter, JRM v5]

The patterns appearing in 6667^2, and similarly in 3334^2 and so on, are examples of a general rule. Any number, of however many digits, will form a pattern when a sufficiently large number of either 3s, 6s or 9s are prefixed to it.

Thus, $72^2 = 5184$; $672^2 = 451,584$ and $6672^2 = 44,515,584$ and so on.

6729
Double 6729 is 13,458, the 2 numbers containing the digits 1 to 9 between them.

6999

When 6999 is reversed and added to itself, $6999 + 9996 = 16,995$, and this process repeated, it takes 20 steps to become a palindrome, and the resulting palindrome is the longest for any number up to 10,000.

7998 also leads after the first step to 16,995.

7140

The largest number that is both triangular and tetrahedral. 7140 is the 119th triangular number and the 34th tetrahedral number.

7314

With 7315, the smallest pair of consecutive numbers, each the product of 4 distinct primes. The next two such pairs are 8294, 8295 and 8645, 8646.

7560

$7560 = 2^3 \times 3^3 \times 5 \times 7$ has 64 factors, more than any other number below 10,000 except for $9240 = 2^2 \times 3 \times 5 \times 7 \times 11$, which also has 64 factors.

7744

$7744 = 88^2$ is the only square with this digit pattern.

8000

$8000 = 20^3$ is the sum of 4 consecutive cubes: $11^3 + 12^3 + 13^3 + 14^3$.

8128

$8128 = 2^6(2^7 - 1)$ is the 4th perfect number.

8169

The number of pairs of twin primes less than 1,000,000. The sequence of numbers of twin primes less than 10^n goes: 2 8 35 205 1224 8169 58,980 440,312 . . . [Sloane 1855]

8191

$8191 = 1 + 90 + 90^2 = 1 + 2 + 2^2 + 2^3 + \ldots + 2^{12}$

$8191 = 2^{13} - 1$ is a Mersenne prime. Note that the index 13 is also prime. It had been conjectured that although most Mersenne numbers appear to be composite, a Mersenne number whose index was a Mersenne prime would itself be prime. This would have provided a formula for an infinite sequence of primes, albeit a sequence that becomes incalculably large very quickly. The conjecture however is false. $2^{8191} - 1$ is composite.

8208

$8208 = 8^4 + 2^4 + 0^4 + 8^4$

8281

$8281 = 91^2$ is a square whose digits form 2 successive integers. This is the only 4-digit square with this property.

8373

With 8433 and 8493, three integers with common difference 60, each of whose 4th power is the sum of 4 other 4th powers. The triplet 8517, 8577, 8637 have the same property. [MOC v27 491]

8712

The subject of one of G. H. Hardy's unserious mathematical properties. It is a multiple of its reversal, 2718. (*See 153.*)

8902

There are 8902 ways of playing the first 4 moves at chess. [Sloane 5100]

9010

$25^2 + 26^2 + 27^2 + \ldots + 623^2 + 624^2 = 9010^2$

9240

It has 64 divisors.

9642

When multiplied by 87,531 it forms the largest product of 2 numbers using the digits 1 to 9 once each.

9801

$9801 = 99^2$ and $98 + 01 = 99$, so 9801 is a Kaprekar number.

9999

$9999^2 = 99980001$ and the two Kaprekar halves, 9998 and 0001, sum to 9999.

Compare $9999^3 = 999700029999$, whose 3 'thirds' sum to 2×9999.

10,001

$10,001 = 73 \times 137$

Compare 101, which is prime, $1001 = 7 \times 11 \times 13$ and $100,001 = 11 \times 9091$.

10,989

$10,989 \times 9 = 98,901$

11,593

This number is the first in a sequence of 9 consecutive primes all of the form $4n + 1$. [Den Haan]

11,826

$11,826^2$ is the smallest pandigital square. It was first noted by John Hill in 1727, who thought it was the only pandigital square.

12,285

Together with 14,595 the smallest pair of odd amicable numbers, discovered by B. H. Brown in 1939.

12,321

$12,321 = 111^2$, both palindromic. So are $1,234,321 = 1111^2$ etc.

12,496

Sociable numbers

12,496 is the first of a chain of 5 sociable numbers, discovered by Poulet in 1918.

The sum of the divisors, excluding itself, of each number is the next number in the chain, the last number preceding the first: 12,496; 14,288; 15,472; 14,536; 14,264; (12,496).

Poulet also found a 28-link chain starting with 14,316. No more were known until 1969 when Henri Cohen found 7 new chains, each of 4 links. Recently, computer searches have found many more, including a 9-link chain starting with 805,984,760. [Ren Yuanhua, Guy 63]

Curiously no chains with just three links have been found, despite diligent searching. There are certainly none with smallest member less than 50 million. Someone named these hypothetical chains 'crowds', so mathematically speaking a crowd is a very elusive phenomenon, and may not exist at all.

12,758

This is the largest number that cannot be represented as the sum of distinct cubes. [Dressler and Parker, MOC v28]

14,316

The start of a remarkable sociable chain of no fewer than 28 numbers, discovered by Poulet in 1918. Starting at the top of the left column, and reading down, the sum of the proper divisors of each number is equal to the next number, 17,716 finally leading back to 14,316:

14,316	629,072	275,444	97,946
19,116	589,786	243,760	48,976
31,704	294,896	376,736	45,946
47,616	358,336	381,028	22,976
83,328	418,904	285,778	22,744
177,792	366,556	152,990	19,916
295,488	274,924	122,410	17,716
			(14,316)

No other sociable chain is known as large, or larger than, this one, despite its venerable age.

16,830

$16,830^3$ is the sum of all the consecutive cubes from 1134^3 to 2133^3. [Beiler]

16,843

Charles Babbage conjectured that $\binom{2p-1}{p-1} - 1$ is divisible by p^2 if and only if p is prime.

The conjecture is false, and the smallest counter-example is $16,843^2$.

Any higher power of 16,843 is also a counter-example. n^2 is not a counter-example for any other n less than 150,000. [David Singmaster]

17,163

This is the largest number that is not the sum of the squares of distinct primes. [Gupta, *Selected Topics in Number Theory*, 1980]

17,296

With 18,416 the second pair of amicable numbers to be discovered.

18,144

The volume of the only rational tetrahedron with edge lengths less than 157. The edges are 117, 80, 53, 52, 51 and 84; the face areas are 1800, 1890, 2016 and 1170. [Buchholz, Guy 191]

18,496

$18{,}496 \times 2^{18{,}496} + 1$ is a Cullen prime, which can also be written in the form $(17 \times 2^{9251})^2 + 1$. [Guy 10]

19,600

Only two numbers are simultaneously square and tetrahedral.

One is the uninteresting $4 = 2^2 = 1 + 3$ and the other is $19{,}600 = 140^2 = 1 + 3 + 6 + 10 + 15 + \ldots + 1176$.

20,161

Every number greater than 20,161 is the sum of 2 abundant numbers.

20,736

12^4 and therefore 10,000 in base 12 or duodecimal.

21,000

The first number to use 3 words in its normal English description: 'twenty-one thousand'.

21,576

The area of the smallest Pythagorean quadrilateral, in which the sides of the four right-angled triangles formed by its edges and its perpendicular diagonals are all integral. The sides are: 25,60,65; 91,60,109; 91,312,325; 25,312,313. [ApSimon, JRM v21 9]

21,952

$21{,}952 = 28^3$. With $64{,}000 = 40^3$, the only two integers whose number of divisors equals their cube root. [JRM v26 156]

25,200

The smallest number which can be written as the product of 4 triples, each with the same sum: $25{,}200 = 6 \times 56 \times 75 = 7 \times 40 \times 90 = 9 \times 28 \times 100 = 12 \times 20 \times 105$. The common sum is 137. [Mauldron, AMM v88]

25,930

$\phi(25{,}930) = \phi(25{,}935) = \phi(25{,}940) = \phi(25{,}942) = 2^7 \times 3^4$.

26,861

primes 4n + 1 and 4n + 3

There are exactly as many primes of the form $4n + 1$ below 26,861 as there are primes of the form $4n + 3$. Since 26,861 is prime of the $4n + 1$ type, it puts the $4n + 1$ primes in the majority, for the first time.

All prime numbers beyond 2 are either of the form $4n + 1$ or $4n + 3$. Which form is most common? The sequence of primes starts like this, where the *italic* type shows primes of the form $4n + 3$: *3* *5* *7* *11* 13 17 *19* *23* 29 *31* 37 41 *43* *47* 53 *59* 61 *67* *71* *73* . . .

Of the first 20 primes, 11, a bare majority, are of the form $4n + 3$. This majority continues however all the way up to 26,849, at which point they are equal in number, and then 26,861 tips the balance, though only momentarily. The next two primes, 26,863 and 26,879 are both '$4n + 3$' types.

Although the $4n + 1$ primes seem to be usually in a minority, Littlewood proved the lead switches from one to the other an infinite number of times. [Bays and Hudson, MOC v32]

27,594
This number can be written in 2 curiously related ways as a product: $27,594 = 73 \times 9 \times 42 = 7 \times 3942$. [Madachy]

28,561
$28,561 = 13^4 = 12^4 + 8^4 + 7^4 + 6^4 + 2^4 + 2^4$.

29,351
Simultaneously a pseudoprime in bases 2, 3, 5 and 7. [Ribenboim 91]

30,031
$30,031 = 2 \times 3 \times 5 \times 7 \times 11 \times 13 + 1 = 59 \times 509$. The first composite number of the form $p\# + 1$.

30,240
The first four-fold perfect number. The sum of its divisors is 120,960. [Sloane 4297]

30,739
For what numbers excluding 6th powers are the decimal parts of the square and cube roots most nearly equal?

Up to 50,000 the smallest difference is found in the square root and cube root of 30,739 whose decimal parts differ by about 0·00001 51.

The 1st integer thereafter to produce a smaller difference is 62,324. The decimal parts of its square and cube roots differ by about 0·00001 1576. [Baumwell and Rubin, JRM v9]

33,614
With 33,615 and 33,616, probably the smallest triplet of successive integers each divisible by a 4th power. [Vandemergel]

33,705
The first exception to the rule that $945 + 630n$ is an odd abundant number. It is deficient. [Madachy quoted in Whallen and Miller, JRM v22 261]

40,311
The start of the longest known sequence of consecutive integers with the same number of divisors: 40,311, 40,312, 40,313, 40,314 and 40,315 each has 8 divisors. [Le Lionnais]

40,320
Factorial 8, or 8!

40,585
Equal to the sum of the factorials of its digits: $40,585 = 4! + 0! + 5! + 8! + 5!$. This was discovered as late as 1964 by Leigh Janes. [Madachy]

40,755
The first number, after 1 and 15, to be simultaneously pentagonal and hexagonal and therefore, also, triangular.

41,041
The smallest Carmichael number with 4 factors. It equals $7 \times 11 \times 13 \times 41$. [Ribenboim 99]

44,488
44,488, 44,489, 44,490, 44,491, 44,492 is the first sequence of 5 consecutive happy numbers.

44,944
$44,944 = 212^2$. Both numbers use only two different digits. [Guy 262]

45,045
Equal to $5 \times 7 \times 3^2 \times 11 \times 13$, and the first odd abundant number to be discovered, by Carolus Bovillus. [Dickson v1 7]

47,619

047619 is the period of 1/21, the smallest number with 2 prime factors that do not divide 10.

The 2 'halves', 047 and 619, sum to 666: it is a multiple of 333 but not of 999: $47,619 = 143 \times 333$.

The 3 'thirds', 04, 76 and 19, sum to 99, so it is a multiple of 99; in fact it is 99×481.

$047619^2 = 2,267,569,161$

Adding the 2 6-figure 'halves': $569161 + 2267 = 571428$, which is the period of 4/7.

50,625

Equal to $15^4 = 4^4 + 6^4 + 8^4 + 9^4 + 14^4$.

This is the smallest example of a 4th power equal to the sum of only 5 distinct 4th powers.

51,984

$51,984 = 228^2 = 37^3 + 11^3$, the smallest square to the sum of 2 cubes, after $3^2 = 1^3 + 2^3$. The next is $671^2 = 65^3 + 56^3$. [Swain, *Mathematical Spectrum* v20 56]

54,748

Equal to the sum of the 5th powers of its digits: $54,748 = 5^5 + 4^5 + 7^5 + 4^5 + 8^5$.

57,321

Together, 57,321 and 60,984 are pandigital, and so are their squares, 3,285,697,041 and 3,719,048,256. There are three other pairs of numbers with this property: 35,172 and 60,984; 58,413 and 96,702; 59,403 and 76,182. [Gardner, *The Incredible Dr Matrix*, 1976]

63,504

$63,504 = 441 \times 144 = 252 \times 252$. The smallest number, not a multiple of 10, to equal the product of a number and its reversal in two ways. Another example, with a palindrome, and therefore a square, is: $7,683,984 = 2772^2 = 1584 \times 4851$. If, however, the palindrome 252 is considered a defect, then a larger non-palindromic example is: $144,648 = 861 \times 168 = 492 \times 294$. [S. S. Gupta]

65,536

2^{16}. A 64K computer memory actually contains 65,536 bytes.

The Ackermann function is one of the fastest increasing functions used in mathematics. Its values from $f(0)$ to $f(5)$ are 1, 3, 4, 8, 65,536.

The only power of 2 up to $2^{31,000}$, inclusive, which contains no digit 1, 2, 4 or 8. [Ashbacher, JRM v22 76]

65,537
Equal to $2^{2^4} + 1$. The 4th Fermat number, and the largest known Fermat prime. It is therefore possible to construct a regular polygon of 65,537 sides by classical methods using a straight edge and compasses only. The 384 quadratics required for an actual construction of the 65,537-gon were calculated by J. G. Hermes in 1894.

69,696
Equal to 264^2 and therefore a palindromic square whose root is not palindromic.

74,162
$74,162^2 = 5,500,002,244$ [JRM v17 84]

076,923
The initial zero is included because this is the decimal expansion of $1/13 = 0.076923\ 076923\ldots$

Multiplied by 3, 4, 9, 10 or 12 the result is a cyclic permutation of the same digits. Multiplied by 2, 5, 6, 7, 8 or 11, the result is a cyclic permutation of 153846.

78,557
Prime numbers of the form $k \times 2^n + 1$ have been much studied, not least because factors of Fermat numbers are always of this form.

The sequence of numbers $78,557 \times 2^n + 1$ is unusual because they are not prime for any positive value of n. Every member of the sequence is divisible by one of the primes 3, 5, 7, 13, 19, 37 or 73.

78,557 is quite possibly the smallest value of k, such that $k \times 2^n + 1$ is always composite. [Baillie, Cormack and Williams, MOC v37]

81,081
The first odd abundant number that does *not* end in 5. The next 5 such numbers are 153,153, 171,171, 189,189, 207,207 and 223,839. [Madachy quoted by Whallen and Miller, JRM v22 259]

87,360

$87,360 = 2^6 \times 3 \times 5 \times 7 \times 13$. A unitary divisor, d, of n is one such that d and n/d have no common factor. A unitary perfect number, n, of which this is an example, is the sum of its unitary divisors, apart from n itself. The only smaller unitary perfects are $6 = 2 \times 3$; $60 = 2^2 \times 3 \times 5$; and $90 = 2 \times 3^2 \times 5$. The 5th unitary perfect is $2^{18} \times 3 \times 5^4 \times 7 \times 11 \times 13 \times 19 \times 37 \times 79 \times 109 \times 157 \times 313$.

Any larger unitary perfect must have at least 8 prime factors and be divisible by a power of 10 greater than 2^{10}. There are no odd unitary perfect numbers. [Guy 53]

90,625

The only 5-digit automorphic number not beginning with a zero. Its square ends in the same digits, . . . 90625.

94,249

Equal to 307^2 and therefore a palindromic square whose root is not itself palindromic.

99,131

There are 35 sets of 4 consecutive primes of the form $10n + 1$, $10n + 3$, $10n + 7$, $10n + 9$, below 100,000. This is the start of the largest. [Ram Nair, 'Numbers Count', *Personal Computer World*, Aug. 1993]

99,954

Kaprekar's process for all 5-digit numbers whose digits are not all equal leads to one of 3 separate cycles. The smallest cycle is $99,954-95,553$. The other two cycles are $98,532-97,443-96,642-97,731$ and $98,622-97,533-96,543-97,641$. [Kordemsky]

100,001

Equal to 11×9091.

All the numbers in the sequence 100,001, 10,000,100,001, 1,000,010,000,100,001, . . . are composite. [Wilke, *Crux Mathematicorum* v5 147]

103,823

$103,823 = 47^3 = 22^2 + 23^2 + \ldots + 67^2 + 68^2$, the smallest representation of a cube as the sum of consecutive squares. The next smallest is 2161^3. [AMM v94 190]

111,777

This is 'the least integer not nameable in fewer than nineteen syllables', and yet it has just been defined (apparently) in eighteen syllables. This is Berry's Paradox. [Russell and Whitehead, *Principia Mathematica to *56*, 1962, p. 61]

142,857

Cyclic numbers

A number beloved of all recreational mathematicians. It is the decimal period of $1/7$: $1/7 = 0.142857\ 142857\ 142\ldots$

$1/7$ is the first decimal reciprocal to have maximum period, that is, the length of its period is only 1 less than the number itself.

Multiplication by any number from 1 to 6 produces a cyclic permutation of the same numbers:

$$142,857 \times 1 = 142,857$$
$$142,857 \times 2 = 285,714$$
$$142,857 \times 3 = 428,571$$
$$142,857 \times 4 = 571,428$$
$$142,857 \times 5 = 714,285$$
$$142,857 \times 6 = 857,142$$

The sequence of digits also makes a striking pattern when the digits are arranged round a circle.

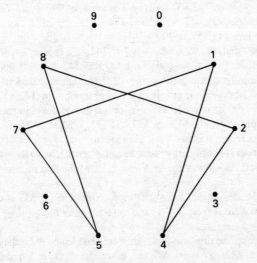

Another pattern is formed by the points (1, 4) (4, 2) (2, 8) (8, 5) (5, 7) and (7,1), which lie on an ellipse. (The points determined by the period of $1/13 = 0.07692\ 30\ \dots$ lie on a hyperbola.) [Kitchen, MM v60 245]

Multiplication by higher numbers produces the same pattern again, with a slight difference. For example, $12 \times 142,857 = 1,714,284$, which becomes 714,285 when the extra 1 is taken from the front and added to the 4 in the units place.

Another example: multiply 142,857 by itself. $142,857^2 = 20,408,122,449$. Separate this number into groups of 6 digits, from the right, and add them: $122,449 + 20,408 = 142,857$.

This makes 142,857 a Kaprekar number.

There is one exception to this pattern: multiplication by 7, or a multiple of 7: $142,857 \times 7 = 999,999$. This is a property of all the periods of repeating decimals. If the period of n is multiplied by n, the result is as many 9s as there are digits in the period of $1/n$. Notice that this relationship is symmetrical. Because $142,857 \times 7 = 999,999$, the decimal period of $1/7$ is 142857 and the decimal period of $1/142,857$ is 7. In fact $1/142,857 = 0.000007\ 000007\ 000007 \dots$

142,857 has another connection with 9. Split 142,857 itself into 2 'halves', and add them: $142 + 857 = 999$. Now any number whose digits when grouped in 3s from the units end add up to 999 is a multiple of 999, and conversely, so 142,857 must be a multiple of 999. Is it? Yes, because 999 divides $999,999 = 7 \times 142,857$ without having any factor in common with 7. In fact $142,857 = 999 \times 143$.

It follows that 999,999, which is $7 \times 142,857$, is also $7 \times 999 \times 143$, and therefore $7 \times 143 = 1001$ and $142,857,143 \times 7 = 1,000,000,001$. This is the basis of a beautiful trick of 'lightning calculation' described by Martin Gardner: to multiply any 9-digit number by 142,857,143, you mentally write the number down twice, so that you would mentally see, for example, 577,831,345 as 577,831,345,577,831,345 and then you simply divide this number by 7. Hey presto!

The answer is doubly impressive because you can write it down, starting from the left, as soon as the first few digits of the 2nd number are given to you.

The 2 'halves' of $1/7$ have another nice property. If 857 is divided by 142 the quotient is 6 $(= 7 - 1)$ and the remainder is 5 $(= 7 - 2)$: $857 = 142 \times 6 + 5$.

If we group the digits in pairs, then for the same reason they will sum to 99: $14 + 28 + 57 = 99$. We can group the digits in 3 and 2, because the period length is 6. Whatever the length of the period, we can 'group' the digits individually. In this case we confirm that 142,857 is divisible by 9: $1 + 4 + 2 + 8 + 5 + 7 = 27$ and $2 + 7 = 9$.

The pattern is actually stronger than that. It is the opposite digits in the circle, as it were, that add to 9. 1 is opposite 8, 4 opposite 5 and 2 opposite 7.

The natural arrangement of the digits on a calculator has a similar symmetry.

It is not unlike the pattern of a magic square, and the numbers have similar properties. 8×1, 7×2 and 4×5 are in arithmetic progression, and so are $4^2 + 5^2$, $2^2 + 7^2$ and $8^2 + 1^2$. [Kaprekar and Khatri]

Adding the odd digits, $1 + 2 + 5 = 8$, while the even digits sum to $4 + 8 + 7 = 19$, and $19 - 8 = 11$. More generally, if the period of $1/p$ is of maximum length, then the sum of the even digits exceeds that of the odd digits by a multiple of 11. [AMM v101 997]

The pattern of adding 'halves' or 'thirds' works for any multiple of 142,857 (with the exception of multiplication by 0), provided the process, as usual, is repeated until 3 digits or 2 digits, respectively, are reached.

$142,857 \times 361 = 51,571,337$: $51 + 571 + 377 = 999$ and $51 + 57 + 13 + 77 = 198$, which becomes 99.

$142,857 \times 74 = 10,571,418$: $10 + 571 + 418 = 999$ and $10 + 57 + 14 + 18 = 99$.

All repeating decimals are effectively geometric series with ratio $1/10$, so it is not surprising that the repeated period of $1/7$ can also be obtained in many ways as a 'diagonal sum', which is equivalent to adding up a geometrical progression. For example, starting from the front:

	or from the back: 7
1	
3	35
9	175
27	875
81	4375
243	21875
729	109375
. . . . .	. . . .
142857 . . .	. . . 857142857142857

Does it seem curious that 142,857 is very nearly 14−28−56 . . . doubling each time? This is also no accident:

```
14
  28
    56
     112
     224
        448
          896
          1792
            . . . .
142857142857142 . . .
```

All of these properties of 142,857 are shared by the periods of any reciprocal whose period is of maximum length, with small adjustments such as the choice of multipliers in the last example.

Numbers of maximum period must be prime but, surprisingly, there is no known method of predicting which primes have maximum period.

17 does; its period is of length 16, and its properties match those of 1/7 very closely. So does 1/19 with period 18, but 1/13 has a period of only 6, so its properties are somewhat more complex.

If the period of $1/n$ is not $n - 1$, it is at least a factor of $n - 1$. The period of 1/13 is 6, one-half of 12.

The first few values of n that produce maximum periods for $1/n$ are 7, 17, 19, 23, 29, 47, 59, 61, and then no more until 97, followed by 109, 113 and 131. This is not many. What proportion of reciprocals of primes have maximum period? About 3/8 according to Shanks, or, if a conjecture of Artin's is correct, 0·37396 . . .

To answer a related question, Shanks has also shown that primes with even decimal periods are exactly twice as numerous on average as primes with odd period length.

What 6-digit number is multiplied by 5 when its unit digit is removed to the front of the number? The answer, of course, is 142,857. This is sometimes called transmultiplication. The problem may just as well ask for the first digit to be placed at the end, or for several digits to be moved in a block. The solution is always the period of some decimal reciprocal.

The reciprocals of composite numbers, such as 21, have more complicated properties. The simplest is that their period is the lowest common multiple of the lengths of the periods of their separate prime factors, if those factors occur singly. $21 = 3 \times 7$, whose reciprocals have periods 1 and 6, so 1/21 has period 6 also. Note that 6 is not a factor of 20.

142,857 is divisible by the repunits 11 and 111.

147,852
Equal to 333×444. The digits 147852 in various orders that are not permutations of the period of 1/7 occur in several other products also. For example, $666 \times 777 = 517,842$ and $333 \times 777 = 258,741$.

148,349
The only number that is equal to the sum of the subfactorials of its digits: $148,349 = !1 + !4 + !8 + !3 + !4 + !9$. [Dougherty]

161,038
The smallest even n, such that $2^n - 2$ is divisible by n, discovered by Lehmer in 1949.

175,560
Equal to $55 \times 56 \times 57 = 19 \times 20 \times 21 \times 22$, the largest of only 3 possible solutions of the equation $x(x + 1)(x + 2) = y(y + 1)(y + 2)(y + 3)$. The others are $x = 2$, $y = 1$, and $x = 4$, $y = 2$. [*Acta Arithmetica* v68 89]

183,184
Equal to 428^2 and therefore a square whose digits form 2 consecutive numbers. There are 3 other 6-digit numbers with the same property: $328,329 = 573^2$, $528,529 = 727^2$ and $715,716 = 846^2$.

196,560
The number of spheres touching any one sphere in a 24-dimensional Leech lattice.

208,335
The largest number that is simultaneously triangular and square pyramidal. It is the 645th triangular number, and the 85th square pyramidal number. [Avanesov, Guy 147]

248,832

Equal to $12^5 = 4^5 + 5^5 + 6^5 + 7^5 + 9^5 + 11^5$.

The smallest representation of a 5th power as the sum of only 6 5th powers.

278,886

Its square starts with a sequence of 5 7s: $278,886^2 = 77,777,400,996$.

314,159

The sequence of prime numbers embedded in the decimal expansion of π goes: 3 31 314,159, then no more until 31,415,926,535,897,932,384, 626,433,832,795,028,841. [Sloane 3129]

333,667

$333,667 \times 296 = 98,765,432$, in which the digits 9 to 2 appear in reverse order. This is the start of a pattern: $33,336,667 \times 2996 = 99,876,654,332$; $3,333,366,667 \times 29,996 = 99,987,666,543,332$, and so on.

The same author shows other patterns involving the same number:

$$333,667 \times 1113 = 371,371,371$$
$$333,336,667 \times 11,133 = 371,137,113,711$$
$$33,333,366,667 \times 1,111,333 = 371,113,711,137,111 \text{ and so on,}$$

or $333,667 \times 2223 = 741,741,741$ and so on.

333,667 also has the property that its square has the digits in ascending order from left to right. This property is shared by any number consisting of a string of 3s, followed by a string of 6s and a solitary 7. For example, $33,366,667^2 = 1,113,334,466,688,889$. The same is true of numbers of these forms: 1666 . . . 6667, 333 . . . 3334 and 333 . . . 3335. [Powers, MM v60 46]

The only prime the decimal period of whose reciprocal is of length 9.

351,120

Its cube can be represented as the sum of 3 cubes, or 4 cubes, or 5 cubes, or 6 cubes, or 7 cubes, or 8 cubes.

362,880

Equal to $9! = 7!3!3!2!$.

369,119

The sum of the primes less than 369,119 is 5,537,154,119, which is divisible by 369,119.

396,733
Together with 396,833, the first pair of consecutive primes that differ by 100.

490,689
$490,689 = 4^3 + 60^3 + 65^3 = 8^3 + 25^3 + 78^3$ and in addition, $4 \times 60 \times 65 = 8 \times 25 \times 78$. [Vandemergel]

509,203
The smallest known value of k, such that $k \times 2^n - 1$ is composite for every value of n. [Keller, Ribenboim 282]

510,510
Equal to the product of the first 7 prime numbers, $2 \times 3 \times 5 \times 7 \times 11 \times 13 \times 17$, and equal to the product of 4 consecutive Fibonacci numbers, $13 \times 21 \times 34 \times 55$, and the product of consecutive integers, 714×715. [Zerger, JRM v12]

523,776
$523,776 = 2^9 \times 3 \times 11 \times 31$

The 3rd tri-perfect number. The sum of its divisors, including itself, is $3 \times 523,776 = 1,571,328$.

548,834
Equal to $5^6 + 4^6 + 8^6 + 8^6 + 3^6 + 4^6$.

621,770
$621,770 + 621,770 = 836^2$ and $621,770 - 077,126$ (its reversal) $= 738^2$. The only other number with this property, less than 10^8 is 65: $65 + 65 = 11^2$ and $65 - 56$ (its reversal) $= 3^2$. [S. S. Gupta]

666,666
To the delight of numerologists, the primitive Pythagorean triangle whose sides are 693, 1924 and 2045 has area 666,666.

698,896
Equal to 836^2. A palindromic square with, a rare event, an even number of digits. Three other palindromic squares with an even number of digits are: $798,644^2 = 637,832,238,736$; and $64,030,648^2$ and $8,316,115,486^2$. [Ondrejka, JRM v20 68]

739,397

The largest 2-sided prime. If digits are removed successively from the left-hand side, the result is a prime; likewise if they are removed successively from the right-hand end (but not both at once).

798,644

The second smallest number whose square is palindromic with an even number of digits: $798,644^2 = 637,832,238,736$.

828,828

A triangular palindrome, whose halves are also palindromic, the only known example, apart from the trivial examples 55 and 66.

872,894

With the next 4 integers, a set of 5 consecutive integers, none of which can be made prime by changing any one digit. [Nelson, *Crux Mathematicorum* v5 147]

1,000,000

$1,000,000 = 10^6$

1,048,576

$1,048,576 = 16^5 = 2^{20}$. Six other permutations of these digits are also perfect squares: 1,056,784, 1,085,764, 5,740,816, 5,764,801, 6,754,801 and 7,845,601. [JRM v19 238]

100,000 in hexademical.

1,122,659

A Cunningham chain of prime numbers is a sequence in which each prime is 1 more than twice the previous member. D. N. Lehmer determined that there were only 3 such chains of 7 primes each with the first member less than 10^7.

The smallest chain is: 1,122,659   2,245,319   4,490,639   8,891,279 17,962,559   35,295,119   71,850,239. [Guy]

1,175,265

Together with 1,438,983, the 1st pair of odd amicable numbers to be discovered, by G. W. Kraft in the seventeenth century.

1,234,321

Equal to 1111^2. Consequently the third line in this pattern:

$$121 \times (1 + 2 + 1) = 22^2$$
$$12,321 \times (1 + 2 + 3 + 2 + 1) = 333^2$$
$$1,234,321 \times (1 + 2 + 3 + 4 + 3 + 2 + 1) = 4444^2$$

and so on.

1,741,725

Equal to $1^7 + 7^7 + 4^7 + 1^7 + 7^7 + 2^7 + 5^7$. There are 3 other numbers equal to the sums of the 7th powers of their digits: 4,210,818, 9,800,817 and 9,926,315. [Gardner, *The Incredible Dr Matrix*, 1976]

1,747,515

Together with 2,185,095 the 3rd pair of triangular numbers whose sum and difference are also triangular.

2,300,000

The earliest inscription in Europe containing a very large number is on the Columna Rostrata, a monument erected in the Roman Forum to commemorate the victory of 260 BC over the Carthaginians. The symbol for 100,000 was repeated 23 times, a total of 2,300,000.

2,759,640

The only values of n for which the quintic $x^5 - x + n$ factors into the product of an irreducible quadratic and an irreducible cubic, are $n = \pm 15$, $\pm 22,440$ and $\pm 2,759,640$. [MM v61 194]

3,424,506

The number of pairs of twin primes less than 1,000,000,000. [Brent, MOC v30 379]

3,628,800

Equal to 10! and the only factorial that is the product of other consecutive factorials apart from the trivial $1! = 0! \times 1!$, $2! = 0! \times 1! \times 2!$ and $1! \times 2! = 2!$.

$\quad$ $10! = 6! \times 7!$

$\quad$ 10! also equals $3! \times 5! \times 7!$.

4,100,625

$4,100,625 = 45^4$. One of only 3 numbers whose 4th roots equal their number of divisors. The other two are $625 = 5^4$ and $6561 = 9^4$. [JRM v26 156]

4,478,976

The smallest known non-trivial solution of the equation $p^p \times q^q = r^r$ is $p = 12^6 = 2,985,984$; $q = 6^8 = 1,679,616$ and $r = 2^{11} \times 3^7 = 4,478,976$. [Le Lionnais]

4,729,494

The cattle problem

4,729,494 occurs as a coefficient in the famous cattle problem attributed to Archimedes. The problem concerns the number of the cattle of the sun, which were divided into 4 herds of different colours, milk white, glossy black, yellow and dappled. 8 conditions then describe the numbers of bulls and cows in each herd. The text is actually ambiguous; it is unclear whether a certain number is to be made square, or merely rectangular. If it has to be a square, then this equation appears: $t^2 - 4,729,494u^2 = 1$.

Such equations are called Pellian, after John Pell, who was thought by Euler to have studied them. There is some evidence that he actually did so, though Euler may well have been mistaking Pell for Lord Brouncker.

Amthor calculated that the least solutions to this equation are:

$t = 109,931,986,732,829,734,979,866,232,821,433,543,901,088,049$
$u = 50,549,485,234,315,033,074,477,819,735,540,408,986,340$

and that in this case the total number of cattle is a number of 206,545 digits, starting 7766 . . . This number has recently been churned out by computer, of course, taking a mere 46 and a bit pages of printout.

It is unlikely that Archimedes could have found such a solution, though he may well have known how to solve this type of equation in principle, and he was interested in very large numbers.

4,937,775

A Smith number, defined by A. Wilansky to be a composite number the sum of whose digits is equal to the sum of the digits of its prime factorization, excluding 1. This was the very first Smith number, named after Wilansky's brother-in-law, H. Smith, whose telephone number this is.

4,937,775 = 3 × 5 × 5 × 65,837 and the digits in each expression sum to 42.

The sequence of Smith numbers starts: 4 22 27 58 85 94 121 166 202 265 . . . [Sloane 3582]

5,134,240

The largest number that cannot be expressed as the sum of distinct 4th powers. [Lin Shen, Guy 137]

5,761,455

The number of primes less than 100,000,000. [Ribenboim 179]

9,843,019

The start of the smallest sequence of 5 consecutive primes in arithmetical progression. The common difference is 30. [Jones, Lal and Blundon, MOC v21 103]

9,999,999

Napier's original logarithms were not 'natural', to base e, nor were they based explicitly on exponents. Napier assigned the number 10,000,000 the logarithm 0, and 9,999,999 the logarithm 1. By multiplying repeatedly by 9,999,999/10,000,000 he constructed a sequence of numbers with logarithms 2, 3 . . . and so on.

In the appendix to the 1618 English translation of Napier's original work there is a table of natural logarithms, probably due to William Oughtred who invented the straight and the circular slide rules.

John Wallis in 1685 and Johann Bernoulli in 1694 realized that logarithms could be thought of as exponents.

10,213,223

A number which, read from left to right, describes itself: it contains one zero, two 1s, three 2s, two 3s. The smallest such sequence is 22. One method of construction is to take a seed sequence, and self-describe it repeatedly, until a self-descriptive string turns up: for example, 231 – 111213 – 411213 – 31121314 – 41122314 – 31221324 – *21322314*.

12,345,679

$$12,345,679 \times 1 = 12,345,679 \quad \text{(digit 8 missing)}$$
$$12,345,679 \times 2 = 24,691,358 \quad \text{(digit 7 missing)}$$
$$12,345,679 \times 3 = 37,037,037$$
$$12,345,679 \times 4 = 49,382,716 \quad \text{(digit 5 missing)}$$
$$12,345,679 \times 5 = 61,728,395 \quad \text{(digit 4 missing)}$$
$$12,345,679 \times 6 = 74,074,074$$
$$12,345,679 \times 7 = 86,419,753 \quad \text{(digit 2 missing)}$$
$$12,345,679 \times 8 = 98,765,432 \quad \text{(digit 1 missing)}$$
$$12,345,679 \times 9 = 111,111,111$$

Note that in each product the sequence 1 to 9, with one digit missing, can be read by cycling through the number, with a suitable repeated jump. For example, 61,728,395 can be read as,

1 2 3 5

and going round again,   6 7 8 9

12,960,000

This is the 2nd Geometric Number of Plato, associated with 216, according to many commentators. It has been derived in various ways, for example as 60^4 or as 4800×2700.

There was a tradition of a Great Year of Plato, though Plato never mentions it, of 36,000 years. At 360 days per year, 36,000 years occupies 12,960,000 days.

12,988,816

The number of ways of tiling a standard chessboard with dominoes. [Sloane 2160]

13,123,110

The area of the smallest triplet of primitive Pythagorean triangles with the same area. The sides are: 7373, 5852, 4485; 3059, 8580, 9109; 1380, 19,019, 19,069. The area of the next such triplet is 2,570,042,985,510. [Shedd, 1945; AMM v99 283]

15,760,091

This prime starts a sequence of consecutive primes with the same pattern of consecutive differences as the sequence 11 13 17 19 23 29 31 37. [Guy 24]

20,615,673

Euler conjectured that a 4th power could not be the sum of 3 4th powers. Noam Elkies made the *New York Times* in 1988 when he discovered this counter-example: $20,615,673^4 = 2,682,440^4 + 15,365,639^4 + 18,796,760^4$. [Elkies, MOC v51 825]

Later, Roger Frye found the smallest solution: $95,800^4 + 217,519^4 + 414,560^4 = 422,481^4$.

23,456,789

A prime with consecutive digits. The sequence of such primes starts: 23 67 89 4567 78,901 678,901 23,456,789 45,678,901 9,012,345,678,901 789,012,345,678,901. It is not known if the sequence can be continued. [JRM v5 254]

24,678,050

Equal to $2^8 + 4^8 + 6^8 + 7^8 + 8^8 + 0^8 + 5^8 + 0^8$.

[182]

33,550,336
Equal to $2^{12}(2^{13} - 1)$.

The 5th perfect number, recorded for the first time anonymously in a medieval manuscript.

33,817,088
$33,817,088 = 2^9 \times 257^2$. The smallest even number n, such that there is no odd number m, such that $\phi(n) = \phi(m)$. [Lorraine Foster, Guy 94]

42,549,416
The smallest number which is the sum of 2 cubes, in 4 ways: $42,549,416 = 348^3 + 74^3 = 282^3 + 272^3 = (-2662)^3 + 2664^3 = (-475)^3 + 531^3$. [AMM v100 336]

50,847,534
The number of primes less than 1,000,000,000. [Ribenboim 179]

60,996,100
Equal to 7810^2 and a square composed of 2 adjacent consecutive numbers, 6099 and 6100.

The only other 8-digit solutions are $9079^2 = 82,428,241$ and $9901^2 = 98,029,801$. [Kraitchik]

73,939,133
The largest prime number that can be 'tailed' again and again by removing its last digit to produce only primes, ending with 739, 73, 7. [M. E. Larsen]

83,623,935
Equal to $3 \times 5 \times 17 \times 353 \times 929$ and with the property that $\phi(n)$ divides $n + 1$. $83,623,935 \times 83,623,937$ has the same property. [Meally, Guy 93]

87,539,319
The smallest number that can be represented as the sum of 2 cubes in 3 different ways:

$87,539,319 = 167^3 + 436^3 = 228^3 + 423^3 = 255^3 + 414^3$. [Leech]

121,174,811
The start of the longest known sequence, of 6 terms, of consecutive primes in arithmetical progression. The common difference is 30. [Ribenboim 226]

123,456,789

When multiplied by 8, it becomes 987,654,312, neatly reversing the last two digits. It also remains pandigital when multiplied by 2, 4, 5 and 7.

There are several numbers that are pandigital, including zero, and remain so when multiplied by several factors. For example, 1,098,765,432 when multiplied by 2, 4, 5 or 7.

129,572,008

One of 2 large repfigits discovered by Clifford Pickover. If a generalized Fibonacci sequence starts with these digits as its first 9 terms, and each subsequent term is the sum of the previous 9, then the sequence includes this number itself. The other such number is 251,133,297. [Pickover, *Computers and the Imagination*, 1991]

139,854,276

Equal to $11,826^2$.

The smallest pandigital square.

160,426,514

The smallest number that can be represented in 2 ways as the sum of 3 6th powers: $160,426,514 = 3^6 + 19^6 + 22^6 = 10^6 + 15^6 + 23^6$. This representation has the additional feature that: $3^2 + 19^2 + 22^2 = 10^2 + 15^2 + 23^2$.

It is known that an infinite number of other solutions exist.

164,597,832

$164,597,832 = 6 \times 7 \times 9 \times 524 \times 831$, both sides being pandigital. [Havard, *Computer Weekly*, 25 Jan. 1996]

207,622,273

The first of 16 consecutive primes of form $4n + 1$. [Vandemergel]

256,103,393

$256,103,393 = 22^4 + 93^4 + 116^4 = 29^4 + 66^4 + 124^4$ and, in addition, $22 \times 93 \times 116 = 29 \times 66 \times 124$. [Vandemergel]

258,474,216

The largest triangular number to be the product of consecutive integers. The others are: 6, 120, 210, 990 and 185,136. [Tzanakis and de Weger, Guy 148]

272,400,600

The sum of the harmonic series, $1 + 1/2 + 1/3 + 1/4 + 1/5 + \ldots$ does tend to infinity, but extremely slowly.

It takes 272,400,600 terms to pass 20. In fact the sum of the first 272,400,599 terms is approximately 19·99999 99979 and adding 1/272,400,600 the total is approximately 20·00000 00016.

It takes 12,367 terms to pass 10, reaching approximately 10·00004 30083.

It takes approximately $1\cdot5 \times 10^{43}$ terms to exceed 100. [Boas and Wrench, AMM v78]

275,305,224

The number of magic squares of order 5, finally calculated in 1973 on computer by Richard Schroeppel. This total excludes rotations and reflections.

363,474,363

The 363,474,363rd triangular number is 66,056,806,460,865,066: both numbers are palindromic. Curiously $T_{363} = 66,066$ has the same property. [Ashbacher, JRM v24 184]

381,654,729

The unique integer such that the number formed by the first n digits is divisible by the digit n. [Upton, *Sunday Times* Brainteaser 1040]

0,429,315,678

This pandigital number is equal to 3 pandigital products: 04,926 × 87,153; 07,923 × 54,186 and 15,846 × 27,093. [Gouffé, JRM v6]

438,579,088

Equal to $4^4 + 3^3 + 8^8 + 5^5 + 7^7 + 9^9 + 0^0 + 8^8 + 8^8$.

The only other number with this property is 3435.

455,052,511

The number of primes less than 10^{10}, calculated by D. N. Lehmer.

The complete table for the number of primes less than powers of 10 up to this figure is:

10	4
10^2	25
10^3	168
10^4	1,229
10^5	9,592
10^6	78,498
10^7	664,579
10^8	5,761,455
10^9	50,847,534
10^{10}	455,052,511

509,033,161

$509,033,161 = (7 \times 13 \times 19)(37 \times 73 \times 109)$. This number and the 2 numbers in the brackets are Carmichael numbers. [Dubner, JRM v22 3]

635,318,657

The smallest known number, discovered by Euler, that can be represented as the sum of 2 4th powers in 2 ways: it is equal to $59^4 + 158^4$ and $133^4 + 134^4$.

923,187,456

The largest pandigital square, if zero is not used. Equal to $30,384^2$.

987,654,321

When multiplied by 1, 2, 4, 5, 7 or 8, the result is pandigital, including the zero.

   Note also: $987,654,321 - 123,456,789 = 864,197,532$.

999,999,937

The largest 9-digit prime, falling short of 10^9 by 63.

1,026,753,849

$1,026,753,849 = 32,043^2$. The smallest pandigital square, including zero.
[Stajsczak]

1,031,223,314

If read in pairs of digits, thus $10-31-22-33-14$, it is self-descriptive. It can be obtained by starting with 10 and then writing down the sequence which describes 10, that is, 1011 (meaning, one-zero and one-one), and then repeating the process to the 10th step. [Mudge, 'Numbers Count', *Personal Computer World*, June 1996]

1,111,111,111
The smallest 10-digit Kaprekar number. Its square is:
1,234,567,900,987,654,321.

1,234,567,891
One of 3 known primes whose digits are in ascending order, beginning with 1 and returning from 9 to 1 or zero where necessary. The other two are: 12,345,678,901,234,567,891 and 1,234,567,891,234,567, 891,234,567,891. [Madachy, JRM, v10]

1,375,298,099
Equal to the sum of 3 5th powers in two ways: $24^5 + 28^5 + 67^5 = 3^5 + 54^5 + 62^5$. [R. Alter]
   It is not known if a number can be the sum of only 2 5th powers in more than one way.

1,480,028,171
Harry Nelson won $100 from Martin Gardner for discovering a 3×3 magic square of primes. This is the central prime. The other cells are this number ± 12, ± 18, ± 30 and ± 42. [Guy 18]

1,533,776,801
The third number to be simultaneously pentagonal and hexagonal and therefore triangular, also.

1,787,109,376
One of only 2 10-digit automorphic numbers, that is, its square ends in the digits . . . 1,787,109,376.
   It follows that any number formed by chopping off digits from the front will also be automorphic. For example, $109,376^2$ ends in the digits . . . 109,376.
   The other 10-digit automorph is 8,212,890,625.

1,857,437,604
A square, $43,098^2$, the sum of whose divisors is a cube, 1729^3. [Beiler]

1,979,339,339
The largest prime, such that chopping off digits from the right-hand end always leaves a prime, counting 1 as a prime.
   Only slightly smaller, with the same property, is 1,979,339,333.

2,438,195,760

This is pandigital and also divisible by every number from 2 to 18. Kordemsky gives 3 other numbers with this property: 4,753,869,120; 3,785,942,160 and 4,876,391,520.

4,294,967,297

The 5th Fermat number, equal to $2^{2^5} + 1$, which Euler showed to be composite, thereby destroying Fermat's conjecture that all numbers of this form are prime. It is equal to $641 \times 6,700,417$.

4,679,307,774

The only 10-digit number equal to the sum of the 10th powers of its digits, discovered by Harry L. Nelson. [Madachy]

4,700,063,497

The smallest number n, for which $2^n - 3$ is divisible by n.

5,391,411,025

$5,391,411,025 = 5^2 \times 7 \times 11 \times 13 \times 17 \times 19 \times 23 \times 29$. This may be the smallest odd abundant number not divisible by 3. [Whallen and Miller, JRM v22 259]

6,661,661,161

$6,661,661,161 = 81,619^2$. The largest known square with only two distinct non-zero digits. [Yoshigahara]

8,549,176,320

The digits in alphabetical order, reading 0 as 'zero'.

9,814,072,356

The largest square, $99,066^2$, that is pandigital, including the zero.

9,876,543,210

Subtract 0,123,456,789 and the answer is 9,753,086,421. All 3 numbers are pandigital with zero.

10,662,526,601

The only known palindromic cube, 2201^3, whose root is not palindromic. There is no other number with this property less than $2 \cdot 8 \times 10^{14}$. [Simmons, JRM v3 93]

The sequence of numbers whose cubes are palindromes starts: 1   2   7   11   101   111   1001   2201   10,001   10,101... All known palindromic

4th powers have palindromic roots. No palindromic 5th powers are known.

12,345,554,321
A palindromic Smith number. [Dudley, MM v67]

14,182,439,040
The smallest 5-fold perfect number. The sum of its divisors is 5 times the number itself. [Guy 45]

15,527,402,881
The smallest 4th power that is the sum of only 4 4th powers: it equals $353^4 = 30^4 + 120^4 + 272^4 + 315^4$.

18,465,126,293
Counting upwards steadily, the number of primes of the form $4n + 3$ exceeds the number of primes of the form $4n + 1$ for most of the first few billion numbers.

The sixth and largest known region for which this is not so stretches from 18,465,126,293 to 19,033,524,538.

36,363,636,364
The square of this number, 1,322,314,049,613,223,140,496, consists of 2 identical 'halves'. [JRM v14]

61,917,364,224
Equal to 144^5, and the sum of 4 5th powers.

64,795,852,800
The common sum of 3 different amicable pairs: 29,912,035,725 and 34,883,817,075; 31,695,652,275 and 33,100,200,525; 32,129,958,525 and 32,665,894,275. [Moews and Moews, MOC v61]

100,895,598,169
Mersenne in a letter to Fermat in 1643 asked for the ratio of $2^{36} \times 3^8 \times 5^5 \times 11 \times 13^2 \times 19 \times 31^2 \times 43 \times 61 \times 83 \times 223 \times 331 \times 379 \times 601 \times 757 \times 1201 \times 7019 \times 823,543 \times 616,318,177 \times 100,895,598,169$ to the sum of its proper divisors.

Fermat replied that the divisors sum to 6 times the original number and that the prime factors of the last number, 100,895,598,169, are 112,303 and 898,423, both of these being prime.

This is a remarkable feat of factorization. Indeed, even Mersenne's request is remarkable given the complete lack of modern calculating aids. Several theories have been put forward to explain how Fermat did it, and subsequent mathematicians have displayed various methods of finding the 2 factors, without reaching any convincing conclusion.

158,753,389,900

There is 1 chance in this number that you will be dealt a complete suit at Bridge.

399,877,410,625

399,877,410,625 ends in the digits 10625 and is the 10,625th square pyramidal number, making it a square pyramorphic number also, the 26th. [Pickover, *Computers and the Imagination*, p. 222]

444,171,597,444

$444,171,597,444 = 666,462^2$, continuing the sequence $22^2 = 484; 212^2 = 44,944$. No 4th term is possible, because no square can end . . . 4444. [Criton]

554,688,278,429

Start of the longest known Cunningham chain, of length 12, which is also a chain of Sophie Germain primes. (A prime q is a Sophie Germain if $2q + 1$ is also a prime.)

608,981,813,029

For all small numbers N, primes of the form $3n + 2$ that are less than N are more numerous than primes of the form $3n + 1$, less than N.

It is known that for an infinity of values of N, the primes $3n + 1$ are in the majority. How large does N have to be for this to happen?

This is the answer, discovered by Carter Bays and Richard H. Hudson on Christmas Day 1976. Was that the only spare computer time they could grab? [MOC v32]

619,737,131,179

The largest number such that any pair of consecutive digits is a prime, and all these primes are different. [*Eureka*, no. 40]

637,832,238,736

A palindromic square with an even number of digits.

1,000,000,000,061

Together with 1,000,000,000,063 an easy to remember pair of twin primes, though by no means the largest known.

1,002,000,000,000

According to Plutarch, Xenocrates made this the number of syllables that could be formed from the letters of the Greek alphabet.

If this story is true, then this is the first recorded attempt to solve a difficult problem in combinations.

6,963,472,309,248

$6,963,472,309,248 = 2421^3 + 19,083^3 = 5436^3 + 18,948^3 = 10,200^3 + 18,072^3 = 13,322^3 + 16,630^3$. The smallest positive integer to be the sum of 2 positive cubes in 4 ways. [Rosenstiel, Dardis and Rosenstiel, 'Numbers Count', *Personal Computer World*, Nov. 1989]

11,410,337,850,553

The smallest term in a sequence of 22 primes in arithmetical progression. The common difference is 4,609,098,694,200. This is the longest such sequence known. [Pritchard, MOC v64 263]

22,222,222,222,222

A Kaprekar number.

46,257,585,588,439

Every square root of an integer has a periodic continued fraction. The square root of this number happens to have the rather long period of 25,679,652. [Williams, MOC v44 523]

277,777,788,888,899

Form a sequence by starting with a number, multiplying its digits together to get the next number and then repeating. The persistence of a number is the number of steps it takes to reach 1. This is the smallest number with persistence 11.

The smallest numbers with persistence 1,2,3 ... 10 are 10, 25, 39, 77, 679, 6788, 68,889, 2,677,889, 26,888,999, 3,778,888,999. There is no number less than 10^{50} with persistence greater than 11. [Sloane 4687]

443,372,888,629,441

$443,372,888,629,441 = 17 \times 31 \times 41 \times 43 \times 89 \times 97 \times 167 \times 331$ is the smallest number n such that for each prime p which divides n, $p^2 - 1$ divides $n - 1$. This is a stronger condition than that for a Carmichael number. [Pinch, MOC v61 381]

0,588,235,294,117,647

The decimal period of $1/17$. Being of maximum length (16 digits including the zero), its properties match those of 1/7. For example, the period of 2/17 starts 117647 . . .

754,788,753,590,897
The largest known (1994) repfigit number of 15 digits. [Shirriff, JRM v26 196]

758,083,947,856,951
Start of longest Cunningham chain, of length 13, of primes of the second kind, $p(n + 1) = 2p(n) - 1$.

5,559,060,566,555,523 [16 digits]
$5,559,060,566,555,523 = 3^{33}$. Half its digits are 5s. Multiply it by 2, 4 or 6 and the answer contains 10 of one digit. Multiply by 8 and 9 out of 17 digits are 4s. [David Roberts]

6,505,941,701,960,039 [17 digits]
The next 863 numbers are composite. This is the first occurrence of a prime gap of 864. [Weintraub, JRM v25]

48,988,659,276,962,496 [17 digits]
$48,988,659,276,962,496 = 38,787^3 + 365,757^3 = 107,839^3 + 362,753^3 = 205,292^3 + 342,952^3 = 221,424^3 + 336,588^3 = 231,518^3 + 331,954^3$

The smallest integer to be the sum of 2 positive cubes in 5 ways. [Dardis, 'Numbers Count', *Personal Computer World*, Feb. 1995]

052,631,578,947,368,421 [17 digits]
The decimal period of 1/19.

It can also be constructed by adding up the powers of 2, 'backwards':

$$
\begin{array}{r}
1 \\
2 \\
4 \\
8 \\
16 \\
32 \\
64 \\
128 \\
256 \\
\cdots \\
\cdots \cdots \, 947368421
\end{array}
$$

Whenever a decimal period is of maximum length, as here, then the periods of the fractions $1/p$, $2/p$, ... up to $(p-1)/p$ can be listed to make a square with equal sums of rows and columns.

1/19	·0	5	2	6	3	1	5	7	8	9	4	7	3	6	8	4	2	1
2/19	·1	0	5	2	6	3	1	5	7	8	9	4	7	3	6	8	4	2
3/19	·1	5	7	8	9	4	7	3	6	8	4	2	1	0	5	2	6	3
4/19	·2	1	0	5	2	6	3	1	5	7	8	9	4	7	3	6	8	4
5/19	·2	6	3	1	5	7	8	9	4	7	3	6	8	4	2	1	0	5
6/19	·3	1	5	7	8	9	4	7	3	6	8	4	2	1	0	5	2	6
7/19	·3	6	8	4	2	1	0	5	2	6	3	1	5	7	8	9	4	7
8/19	·4	2	1	0	5	2	6	3	1	5	7	8	9	4	7	3	6	8
9/19	·4	7	3	6	8	4	2	1	0	5	2	6	3	1	5	7	8	9
10/19	·5	2	6	3	1	5	7	8	9	4	7	3	6	8	4	2	1	0
11/19	·5	7	8	9	4	7	3	6	8	4	2	1	0	5	2	6	3	1
12/19	·6	3	1	5	7	8	9	4	7	3	6	8	4	2	1	0	5	2
13/19	·6	8	4	2	1	0	5	2	6	3	1	5	7	8	9	4	7	3
14/19	·7	3	6	8	4	2	1	0	5	2	6	3	1	5	7	8	9	4
15/19	·7	8	9	4	7	3	6	8	4	2	1	0	5	2	6	3	1	5
16/19	·8	4	2	1	0	5	2	6	3	1	5	7	8	9	4	7	3	6
17/19	·8	9	4	7	3	6	8	4	2	1	0	5	2	6	3	1	5	7
18/19	·9	4	7	3	6	8	4	2	1	0	5	2	6	3	1	5	7	8

1/19 has the curious feature of summing to the same constant, 81, along both diagonals as well, and therefore is truly magic. [Andrews, *Magic Squares and Cubes*, 1960]

62,638,280,004,239,857 [17 digits]

The first term of a generalized Fibonacci sequence, in which each term is the sum of the previous two, in which every term is composite. The second term is 49,463,435,743,205,655. [Knuth, MM v63 21]

$2^{58} + 1$ [18 digits]

This number was factorized by Landry in 1869. He commented:

No one of our numerous factorizations of the numbers $2^n \pm 1$ gave us as much trouble and labour as that of $2^{58} + 1$. This number is divisible by 5 and if we remove this factor we obtain a number of 17 digits whose factors have 9 digits each. If we lose this result we shall miss patience and courage to repeat all calculations we have made and it is possible that many years will pass before someone else will discover the factorization of $2^{58} + 1$.

Less than 10 years later, Aurifeuille pointed out that $2^{58} + 1$ can be factorized algebraically as $(2^{29} - 2^{15} + 1) (2^{29} + 2^{15} + 1)$.

Lucas generalized this result to:

$$2^{4n+2} + 1 = (2^{2n+1} - 2^{n+1} + 1) (2^{2n+1} + 2^{n+1} + 1)$$

Such factorizations are now called Aurifeuillian. [*Scripta Mathematica*, v128]

1,111,111,111,111,111,111 [19 digits]
Repunits

A number whose digits are all units was named a 'repunit', short for repeated unit, by Albert Beiler. The name is further abbreviated to R_n where n is the number of units.

Thus $R_1 = 1$, and $R_2 = 11$, the smallest prime repunit. The second smallest prime repunit is R_{19}, which is the number of this entry, as a careful count of the 1s will confirm. It was discovered in 1918 by one of the readers of H. E. Dudeney's newspaper puzzle column.

The only other known primes R_{23} and R_{317} and (almost certainly) R_{1031}. R_{317} was found by H. C. Williams in 1978, after John Brillhart had mistakenly announced that it was composite.

Williams is currently working on the proof that R_{1031} is prime, after probabilistic tests have shown that it was 'almost certainly' prime, without providing the complete certainty that mathematicians desire.

Repunits have a simple relationship to powers of 10: $R_n = (10^n - 1)/9$. For this reason the problem of discovering which repunits are prime and if possible factorizing the others is similar to the problem of Mersenne numbers of the form $2^n - 1$.

The first such table was published by William Shanks, the calculator of π, in 1874 as an aid to finding a prime from the length of the period of its decimal reciprocal.

All the repunits from R_1 to R_{66} have been completely factored into primes. R_{67} and R_{79} are the first repunits whose factorization is still undecided. R_{67} is divisible by 493,121 and R_{79} has factors $317 \times 6163 \times 10,271 \times 307,627$.

Some curiously patterned primes appear as factors.

$R_{38} = 11 \times 909,090,909,090,909,091 \times 1,111,111,111,111,111,111$

This pattern arises very simply. $38 = 2 \times 19$, so:

$10,000,000,000,000,000,001 \times 1,111,111,111,111,111,111 = R_{38}$

and since 19 is an odd number,
10,000,000,000,000,000,001 = 11 × 909,090,909,090,909,091

This pattern, or patterns like it, can be used to break down all even-indexed repunits. The question is, are these giant factors themselves prime? It depends on the number of digits. The midget 9091 is a factor of R_{10} and 909,091 divides R_{14} and 909,090,909,090,909,090,909,090,909,091 divides R_{62}.

For a variation, 9901 divides R_{12}, and both 9901 and 99,990,001 divide R_{24}, while R_{39} is divided by 900,900,900,900,990,990,990,991.

The connection with decimal reciprocals lies in the fact that since, to take an example, the period of 1/7 is 142857, so therefore, 142,857 × 7 = 999,999 = 9 × 111,111 = 9 × R_6.

Working backwards from the table, 7 is a factor of R_6; its period is therefore 9 times the product of the other factors = 9 × 3 × 11 × 13 × 37 = 142,857.

There are many surprisingly large primes whose reciprocals have relatively very short periods. The period of 4649, without the initial zeros, is only 3 × 3 × 239 = 2151. Since 4649 divides R_7, 1/4649 is actually 0·00021 51000 2151 . . .

Repunits have many other properties. Repunits are never squares. It is not known if any repunits other than 1 are cubes, or for that matter if there exists an infinite number of repunit primes.

R_p and R_q are coprime if and only if p and q are coprime.

In base 9, every repunit is also triangular. [G. W. Wishard]

The squares of repunits make a pretty pattern:

For example: $1111^2 = 1,234,321$
and $1,111,111,111^2 = 12,345,678,900,987,654,321$

J. A. H. Hunter mentions that the next square to end in the same 10 digits is:

    $2,380,642,361^2 = 5,667,458,050,987,654,321$

18,446,744,073,709,551,615   [20 digits]
Equal to $2^{64} - 1$.

According to an old legend, Sissa ben Dahir was offered a reward by the Indian King Shirham for inventing the game of chess. Sissa cunningly replied, as recounted by Kasner and Newman:

'Majesty, give me a grain of wheat to place on the first square, and two grains of wheat to place on the second square, and four grains of wheat to place on the third, and eight grains of wheat to place on the fourth, and so, O King, let me cover each of the 64 squares of the board.'

'And is that all you wish, Sissa, you fool?' exclaimed the astonished King.

'Oh, Sire,' Sissa replied, 'I have asked for more wheat than you have in your entire kingdom, nay, for more wheat than there is in the whole world, verily, for enough to cover the whole surface of the earth to the depth of the twentieth part of a cubit.' [*Mathematics and the Imagination*, 1959]

One twentieth part of a cubit is about an inch. Sisa asked for a total of $2^{64} - 1$ grains of wheat. It just so happens that this is the same number of moves required by the priests of the temple at Benares to transfer the 64 golden discs in the thoroughly spurious legend created round the Tower of Hanoi puzzle.

26,072,323,311,568,661,931   [20 digits]
With   43,744,839,742,282,591,947,   118,132,654,413,675,138,222,
186,378,732,807,587,076,747   and   519,650,114,814,905,002,347,   a quintuple of numbers, the sum of any 3 of which is a square. [Wagon, MOC v64 1755]

43,252,003,274,489,856,000   [20 digits]
Equal to $\dfrac{8! \times 12! \times 3^8 \times 2^{12}}{2 \times 3 \times 2}$.

This is the total number of positions that can be reached on the original $3 \times 3 \times 3$ Rubik's Cube.

109,418,989,131,512,359,209   [21 digits]
Equal to 9^{21}. This is the largest *n*-digit number that is also an *n*th power. [Friedlander]

112,359,550,561,797,732,809   [21 digits]
The smallest number which, when 1 is placed at both ends, the number is multiplied by 99. [Pomerance and Hunsucker, FQ v13]

$2^{67} - 1$ [21 digits]
The 67th Mersenne number, which Mersenne claimed was prime. F. N. Cole proved in 1903 that it is composite.
   As E. T. Bell recalls:

At the October, 1903, meeting in New York of the American Mathematical Society, Cole had a paper on the programme with the modest title, 'On the Factorization of Large Numbers'. When the chairman called on him for his paper, Cole – who was always a man of very few words – walked to the board and, saying nothing, proceeded to chalk up the arithmetic for raising 2 to its 67th power. Then he carefully subtracted 1. Without a word he moved over to a clear space on the board and multiplied out, by longhand:
   193,707,721 × 761,838,257,287

The two calculations agreed . . . For the first and only time on record, an audience of the American Mathematical Society vigorously applauded the author of a paper delivered before it. Cole took his seat without uttering a word. Nobody asked him a question. [*Mathematics: Queen and Servant of Science*, 1952]

Bell later asked him how long it had taken him to find this factorization. Cole replied, 'Three years of Sundays.'

154,345,556,085,770,649,600   [21 digits]
The smallest 6-fold perfect number. The sum of its divisors is 6 times the number itself.

0,434,782,608,695,652,173,913   [21 digits]
The period of $1/23$, of maximum length.

11,111,111,111,111,111,111,111   [23 digits]
The 23rd repunit, and only the 3rd prime repunit.

17,796,126,877,482,329,126,044   [23 digits]
The first in a sequence of 9 consecutive integers, each with 9 divisors. [Guy 74]

357,686,312,646,216,567,629,137   [24 digits]
The largest prime number in base 10 such that if you behead it again and again the resulting numbers are all prime, ending with the sequence 9137, 137, 37, 7.

(0 is excluded as a leading digit, because there are almost certainly indefinitely large primes of the form $10^n + 3$, for example.) [Angell and Godwin, MOC v31]

3,608,528,850,368,400,786,036,725   [25 digits]
This has the property that the number formed by the first n digits is divisible by n. Thus, 3,608,528 is divisible by 7, and so on. There is no 26-digit number with this property. [Malcolm Lines, *A Number for Your Thoughts*]

244,197,000,982,499,715,087,866,346   [27 digits]
A large Guiga number, equal to $2 \times 3 \times 11 \times 23 \times 31 \times 47{,}137 \times 28{,}282{,}147 \times 3{,}892{,}535{,}183$. [AMM v103 45]

2,235,197,406,895,366,368,301,560,000   [28 digits]
The reciprocal of the probability that all 4 players at Bridge will be dealt a complete suit. As Martin Gardner points out forcefully, claims that all 4 players have received a complete suit are far more commonly heard

than claims that 2 players have done so, although the latter is far, far more probable.

10^{33} [34 digits]

With 10^{18}, the only powers of 10 between 10^9 and 10^{5000} which can be expressed as the product of 2 zero-free factors:

$$10^{33} = 8,589,934,592 \times 116,415,321,826,934,814,453,125$$

[Mike Mudge, 'Number Count', *Personal Computer World*, Sept. 1985]

1,786,772,701,928,802,632,268,715,130,455,793 [34 digits]

Together with 1,059,683,225,053,915,111,058,165,141,686,995, the start of a generalized Fibonacci sequence (in which each term is the sum of the previous two) in which every member is composite although the first 2 terms have no common factor.

115,132,219,018,763,992,565,095,597,973,971,522,401 [39 digits]

The largest known pluperfect digital invariant in base 10. It is equal to the sum of the 39th powers of its digits.

$2^{127} - 1$ [39 digits]

The 127th Mersenne number.

Lucas, using new methods, announced in 1876 that $M_{127} = 170,141,183,460,469,231,731,687,303,715,884,105,727$ is prime. He later expressed some doubt about this result but it was confirmed in 1914 by Fauquembergue.

This number held the record as the largest known prime of any kind longer than any other, from 1876 to 1951. It was also the largest prime to be discovered without the help of modern calculating aids.

191,918,080,818,091,909,090,909,190,818,080,819,191 [39 digits]

A Sophie Germain prime, because, if this number is N, then $2N + 1$ is also prime. It is also palindromic. As it happens, $2N + 1$ is also palindromic, and a Sophie Germain prime, and its companion, $2(2N + 1) + 1$, is also palindromic. [Dubner, JRM v26 41]

69,720,375,229,712,477,164,533,808,935,312,303,556,800

[41 digits]

This number is divisible by every number up to and including 100. [David Roberts]

10,112,359,550,561,797,752,808,988,764,044,943,820,224,719

[44 digits]

The unique number which, when the last digit, 9, is moved to the front, is the number multiplied by 9.

802,359,150,003,121,605,557,551,380,867,519,560,344,356,971

[45 digits]

The first number in the largest known sequence of primes of the form p, $p + 2, p + 6, p + 8$. [JRM v14]

374,144,419,156,711,147,060,143,317,175,368,453,031,918,731,001,856

[51 digits]

$= 2^{168}$. It contains no digit 2. [David Roberts]

10^{51} [52 digits]

The Sandreckoner

Archimedes in his book *The Sandreckoner*, which he addressed to Gelon, King of Syracuse, describes his own system of counting immense numbers. He starts with the myriad, which was 10,000, and counts up to a myriad myriads describing these as numbers of the first order. He then takes 1 myriad myriad, or 100,000,000 in our notation, to be the unit of the numbers of the second order . . . and he continues until he reaches the myriad-myriadth order of numbers. Archimedes is by no means finished! All the numbers constructed so far are only the numbers of the first period! He continues on his gigantic construction until he reaches 'a myriad-myriad units of the myriad-myriadth order of the myriad-myriadth period'. The highest number in his notation would now be expressed as $10^{80,000,000,000,000,000}$.

He next proposed to count not merely the number of grains of sand on a seashore, or in the whole earth, but the number of grains of sand required to fill the entire universe. Assuming that one poppy-head would contain not more than 10,000 grains of sand, and that its diameter is not less than 1/40 of a finger's breadth, and assuming that the sphere of the fixed stars, which was to Archimedes the boundary of the universe, was less than 10^7 times the sphere exactly containing the orbit of the sun as a great circle . . . the number of grains of sand required to fill the universe turns out to be, in our notation, less than 10^{51}.

By comparison, Edward Kasner and James Newman in discussing a googol, 10^{100}, estimate the number of grains of sand on Coney Island at 10^{20}.

This extraordinary achievement by Archimedes is unique within Greek mathematics. The Greeks generally had no interest in numbers outside of some geometrical context. However, to the east, Indian Buddhist mathematicians did construct immense 'towers' of numbers, rising in multiples of 10 or 100, in order to count the atoms 'even in the 3 thousand thousand worlds contained in the universe'. Perhaps Archimedes was inspired by these Indian achievements to construct his own system. [Heath, *Works of Archimedes*, Dover]

808,017,424,794,512,875,886,459,904,961,710,757,005,754,368,000, 000,000 [54 digits]

$= 2^{46} \times 3^{20} \times 5^9 \times 7^6 \times 11^2 \times 13^3 \times 17 \times 19 \times 23 \times 29 \times 31 \times 41 \times 47 \times 59 \times 71.$

The order of the Monster sporadic simple group, discovered by Fischer in 1974.

$3^9 7^3 11^3 13^3 17^3 41^3 43^3 47^3 443^3 499^3 3583^3$ [59 digits]

The smallest cube whose sum of divisors is also a perfect cube. [Rubin, JRM v27 229]

10^{63} [64 digits]

A vigintillion in American-English words (vigillion according to one author) and, according to several authors, the largest number considered by Archimedes in *The Sandreckoner*.

The extra '3' arises because a million is 10^6 but a billion is only 10^9 and vigintillion is therefore 10 to the power $(3 \times 20) + 3$.

Similarly, centillion is 10^{303}, and by suitable combinations of Latin-sounding words, even larger powers of 10 can be expressed. 10^{366}, for example, is primo-vigesimo-centillions and – wait for it! – milli-millillion is $10^{3,000,003}$.

Milli-millillion may well be one of the least frequently used words in the English language. As the author of the article I am quoting forlornly comments, 'Names for these larger numbers have been so little needed that one can find few places where they have been written.'

In Jaina works *c.* 100 BC *koti* was a hundred-hundred-thousand, one hundred-hundred-thousand koti was called *pakoti* and so on up to *asankhyeya*, which we would represent as 10^{140}.

$2^4 \times 7 \times 9,288,811,670,405,087 \times 145,135,534,866,431 \times 313,887, 523,966,328,699,903$ [65 digits]

Together with its amicable friend, $2^4 \times 7 \times 9,288,811,670,405,087 \times 45,556,233,678,753,109,045,286,896,851,222,527$, the largest known pair of amicable numbers. [Yan and Jackson, *Computer Mathematics and Applications* v27]

$2^{229} - 1$ [69 digits]

Euler proved that $2^{31} - 1$ is prime. The Mersenne numbers from M_{32} to M_{257}, the highest value claimed to be prime by Mersenne, were not finally checked until 1947, when H. S. Uhler, using a desk calculator, finally proved that all of M_{157}, M_{167}, M_{193}, M_{199}, M_{227} and this number, M_{229}, were composite.

In fact the next Mersenne prime does not appear until $2^{521} - 1$. As Uhler remarked, he had no idea when he began his labour that the next Mersenne prime would be so far away!

$2^{257} - 1$ [78 digits]

Mersenne had conjectured that this number is prime. M. Kraitchik showed in 1922 that it is composite, without finding any actual factor. It has since been factored by Baillie and Penk. It equals: 536,006,138,814,359 × 1,155,685,395,246,619,182,673,033 × 374,550,598,501,810,936,581, 776,630,096,313,181,393.

'..........'   [100 digits]

Factorizing large random numbers

How large a number can be chosen at random and factorized within a reasonable time, say a matter of hours, or at the most, days? In 1659 Johann Rahn published a table of factors of numbers up to 24,000. J. P. Kulik (1773–1863) spent 20 years of his life compiling a table of factors up to 100,000,000, a mere 8 digits.

Every extra digit means roughly 10 times as many numbers to consider. With every extra digit the time and effort needed to find the factor of a number without some special pattern multiplies. Only numbers of special forms, such as Mersenne numbers, $2^n - 1$, or Fermat numbers $2^{2^n} + 1$, can be tested to much higher limits. As late as 1943 an author wrote, '. . . in the case of numbers with fifteen digits or more the test of a number's primeness would require years even if we employ all the known methods which facilitate such an examination.'

The known methods included desk calculators, and Lehmer's electro-mechanical sieve, but no electronic computers. By 1974 powerful computers were readily available, and more powerful tests, and it was possible easily to test numbers of 20 to 25 digits.

In 1980 Adleman and Rumely developed a test that would decide if a randomly chosen number of up to 100 digits was prime in 4–12 hours with a large computer. This has been improved by Cohen and Lenstra to run about 1000 times faster. It can now test a 100-digit number in about 40 seconds on a supercomputer, such as the Control Data Cyber 170-750 or the CRAY.

The problem of factoring very large numbers became a matter of public interest and military concern when, in 1975, Whitfield Diffie and Martin Hellman invented the trapdoor function, and shortly afterwards Rivest, Shamir and Adleman showed how to make it a practical proposition. This is a mathematical function that will change any number A into its code number B. The function also has an inverse, which can be used to calculate A from B. The beauty of their idea lies in the relationship between these functions. The inverse could not, in practice, be calculated from the original function.

The heart of the simplest of these functions is a number that is the product of 2 large primes. Rivest's example is two 63-digit primes. These are multiplied together to create a number of 125 or 126 digits. All the enemy agent has to do in order to read Mata Hari's private correspondence is to take the 125/6-digit number and turn it back into the product of the 63-digit numbers. Rivest estimated in 1977 that this would take on a powerful computer about 4×10^{16} years.

The problem of factorizing very large numbers is only one of many techniques based on what are called NP-problems, which share a common feature. They each come in different sizes, depending on the number of digits to be factored, or the size of the knapsack to be filled exactly, and the time taken to solve them by computer increases rapidly as the size of the problem increases.

As a result of work such as Cohen's and Lenstra's, the prime numbers in future will have to be a little longer.

$(11^{104} + 1)/(11^8 + 1)$ [100 digits]
Equal to 86,759,222,313,428,390,812,218,077,095,850,708,048,977 × 1,084,881,048,536,374,706,129,613,998,429,729,484,098,346,115,257, 905,772,116,753.

This was the first 100-digit number to be factorized using a general-purpose factorization algorithm, which did not exploit any particular properties of the original number, or use probabilistic methods. Lenstra and Manasse were responsible. The date was 12 October 1988. [Ribenboim 478]

10^{100}   [101 digits]
The googol
As Edward Kasner and James Newman describe it, in *Mathematics and the Imagination*:

A googol is this number which one of the children in the kindergarten wrote on the blackboard:
10000000000000000000000000000000000000000000000000000000000000000000 000000000000000000000000000000000000000

The definition of a googol is: 1 followed by a hundred zeros. It was decided, after careful mathematical researches in the kindergarten, that the number of raindrops falling on New York in 24 hours, or even in a year or in a century, is much less than a googol.

The child, Dr Kasner's 9-year-old nephew, suggested the name *googol* for this number and *googolplex* for a still larger number, which it was agreed would be 1 followed by a googol of zeros, or 10^{googol}.

The authors write with foresight, from a quarter of a century ago, that such a number might be of real use in combinatorial problems.

In contrast, the total number of particles in the universe has been variously estimated at numbers from 10^{80} up to 10^{87}.

114,381,625,757,888,867,669,235,779,976,146,612,010,218,296,721,242, 362,562,561,842,935,706,935,245,733,897,830,597,123,563,958,705,058, 989,075,147,599,290,026,879,543,541   [129 digits]

Rivest, Shamir and Adelman first thought of using a very large prime number as a public key to a cryptography system. To prove the power of their idea, 'they challenged the world to find the two factors of this 129-digit number, known to people in the field as R S A 129 . . . They were sure that a message they had encrypted using the number as the public key would be totally secure forever. But . . . In 1993 a group of more than 600 academics and hobbyists from around the world began a methodical assault on the 129-digit number, using the Internet to coordinate the work of various computers. In less than a year they factored the number into two primes, one 64 digits long and the other 65 . . . The encoded message says, "The magic words are squeamish and ossifrage." '
[Bill Gates, *The Road Ahead*, 1995, p. 109]

Subsequently, on 12 April 1996, R S A 130, with 130 digits, 18070 . . . 80557, was factorized by a Dutch team into two prime factors, each with 65 digits.

$2^{2^9} + 1 = 2^{512} + 1$ [155 digits]

The 9th Fermat number. The smallest prime factor is 2,424,833. [Lenstra and Manasse 1990]

82,818,079,787,776,757,473,727,170 . . . 10,987,654,321
                [155 digits]

This is the only prime number formed by starting with 100 or less, and writing the sequence of natural numbers back to 1. [Nicol and Filaseta]

$2^{521} - 1$ [157 digits]

The 13th Mersenne prime, leading to the 13th perfect number.

In a few hours on the night of 30 January 1952, using the SWAC computer, Lehmer proved that $2^{521} - 1$ and the 183-digit number $2^{607} - 1$ are both Mersenne primes.

Lehmer used a theorem based on one of Lucas's ideas. Construct the sequence: $4\ 14\ 194\ 37,634\ldots$ in which $s(1) = 4$ and $s(n + 1) = s(n)^2 - 2$. Then the pth Mersenne number is prime if and only if it divides $s(p - 1)$.

11,111,111, ... 111,111 [317 unit digits]

The 3rd known prime repunit.

$(10987654321234567890)_{42}1$ [841 digits]

The $()_{42}$ notation indicates that the bracketed sequence is repeated 42 times, followed by a single digit 1. This is the largest known alternating digit (even–odd) palindromic prime. [Dubner, JRM v26 256]

450! [1001 digits]

Horace Uhler in the 1950s calculated the value of 450! without the aid of an electronic computer, found that it had exactly 1001 digits, and so named it the Arabian Nights Factorial.

$10^{1000} + 81,918 \times 10^{498} + 1$ [1001 digits]

The smallest titanic (more than 1000 digits) palindromic prime. [Dubner, JRM v26 256]

1111 ... 111111 [1031 digits]

The largest known prime repunit, discovered by Williams and Dubner in 1986. It is also the only prime whose decimal period is 1031. [Williams and Dubner, MOC v47 703]

$7532 \times (10^{1104} - 1)/(10^4 - 1) + 1$ [1104 digits]

The largest known prime all of whose digits are prime numbers, discovered by Dubner in 1988. [Ribenboim 477]

$10^{641} \times (10^{640} - 1)/9 + 1$ [1281 digits]

Discovered by Dubner in 1984. The largest known prime with digits all equal to 0 or 1. [Ribenboim 478]

$39,051 \times 2^{6001} - 1$ [1811 digits]

The largest known Sophie Germain prime. [Ribenboim 79]

$2^{4253} - 1$ [1281 digits]

The 19th Mersenne prime, and the first known prime to have more than 1000 digits. It was discovered by Hurwitz in 1961 using an IBM 7090.

$2^{8191} - 1$ [2466 digits]

M_{8191}, the 8191st Mersenne number.

Catalan conjectured that if p is a Mersenne prime, then M_p will be prime. M_3, M_7, M_{31} and M_{127} are indeed prime, but although $8191 = M_{13}$, M_{8191} is composite. This was proved by Wheeler in 1953 on the ILLIAC, in one hundred hours.

Conjectures of this kind seem among the easiest to make in mathematics, and the least likely to be successful.

$2 \times 10^{3020} - 1$ [3021 digits]

The largest known prime with one 1 and all other digits 9; discovered by Williams in 1985. [Ribenboim 477]

$1_{111}2_{111}3_{111} \ldots 8_{111}9_{111}0_{2284}1$ [3284 digits]

The notation indicates that the digits 1 to 9 are each repeated 111 times, followed by 2284 zeros and a 1. The number is prime. [Dubner, JRM v18 89]

$2^{11,213} - 1$ [3376 digits]

The 23rd Mersenne prime, discovered by Gillies at the University of Illinois in 1963. The University celebrated by franking its letters with a special postmark.

$1358 \times 10^{3821} - 1$ [3825 digits]

The largest known prime all of whose digits are odd, discovered by Dubner in 1988. [Ribenboim 477]

$1477! + 1$ [4042 digits]

The largest known prime of the form $n! + 1$, which is also prime for $n = 1, 2, 3, 11, 27, 37, 41, 73, 77, 116, 154, 320, 340, 399, 427, 872$. There are no more for ≤ 4580. [MOC v64 889]

$190{,}116 \times 3003 \times 10^{5120} - 1$ and $190{,}116 \times 3003 \times 10^{5120} + 1$

[5129 digits]

The largest known pair of twin primes, found by Harvey Dubner on 5 October 1995, after a one-day search. ['Numbers Count', *Personal Computer World*, Feb. 1996]

$2^{19{,}937} - 1$

[6002 digits]

The 24th Mersenne prime, discovered by Bryant Tuckerman in 1971.

$2^{21{,}701} - 1$

[6533 digits]

The 25th Mersenne prime, discovered to the delight and amazement of the American public by two 18-year-old school students, Laura Nickel and Curt Noll, in 1978.

$15{,}877\# - 1$

[6845 digits]

The largest prime of the form $p\# - 1$, which is also prime for $p = 3, 5, 11, 41, 89, 317, 337, 991, 1873, 2053, 2377, 4093, 4297, 4583, 6569, 13{,}033$. There are no more with $p \leqslant 35{,}000$. [Caldwell, MOC v64]

$2^{23{,}209} - 1$

[6987 digits]

The 26th Mersenne prime, discovered by Curt Noll in 1979 using the same CDC-CYBER-174 computer on which he and Laura Nickel had found the previous record prime. It took Noll more than 8 hours to check this number on the CDC. Two weeks later David Slowinski used a CRAY-1 supercomputer to check Noll's result, and the calculation was all over in 7 minutes! [Devlin, *Microchip Mathematics*, 1984]

$24{,}029\# + 1$

[10,387 digits]

The largest prime of the form $p\# + 1$, which is also prime for $p = 2, 3, 5, 7, 11, 31, 379, 1019, 1021, 2657, 3229, 4547, 4787, 11{,}549, 13{,}649, 18{,}523, 23{,}801$. There are no more with $p \leqslant 35{,}000$. [Caldwell, MOC v64]

$3610! - 1$

[11,277 digits]

The largest prime of the form $n! - 1$, which is also prime for $n = 3, 4, 6, 7, 12, 14, 30, 32, 33, 38, 94, 166, 324, 379, 469, 546, 974, 1963, 3507$. There are no more with $n \leqslant 4580$. [Caldwell, MOC v64]

$10_{5901}14656410_{5901}1$

[11,811 digits]

The subscript notation indicates the number of times that the preceding digit is repeated. This is the largest known palindromic prime. [Dubner, JRM v26 256]

$2^{44,497} - 1$ [13,395 digits]

The 27th Mersenne prime, discovered by Harry Nelson and David Slowinski using the CRAY-1 supercomputer in April 1979.

$8423 \times 2^{59,877} + 1$ [18,029 digits]

The largest known prime of the form $k \times 2^n + 1$, discovered by Buell and Young in 1988. [Ribenboim 281]

$2^{65,536}$ [19,729 digits]

Equal to $2^{2^{2^{2^2}}}$.

Several functions have occurred quite naturally in recent years in combinatorial problems, which grow astonishingly quickly.

Ackermann's function is defined by $f(a,b) = f((a - 1), f(a, b - 1))$ where $f(1, b) = 2b$ and $f(a, 1) = a$ for a greater than 1.

$f(3, 4) = 2^{65,536}$, which has more than 19,000 digits. Try to imagine the size of $f(10, 10)$ let alone $f(100, 100)$.

As R. L. Graham says of another exploding function, 'It is hard to grasp how fast it grows. It grows so quickly that the numbers somehow begin to lose their meaning.'

[Gina Kolata, 'Does Gödel's Theorem Matter to Mathematics?', *Science*, 218]

$2^{86,243} - 1$ [25,962 digits]

Probably, but not certainly, the 28th Mersenne prime, hunted down by David Slowinski on his trusty CRAY-1 in 1983.

$3!^{3!^{3!}}$ [36,305 digits]

Equal to $6^{46,656}$. This is called the superfactorial of 3. The previous two superfactorials are uninteresting, merely 1! and $2!^{2!}$.

$2^{132,049} - 1$ [39,751 digits]

The 30th Mersenne prime, discovered by David Slowinski using two CRAY-1 computers linked together, on 19 September 1983.

$2^{859,433} - 1$ [258,716 digits]

The third largest known prime. Announced 4 January 1994 by David Slowinski on the Internet. Found by him and Paul Gage. It is the 33rd Mersenne prime to be found, though they may be smaller Mersenne primes between this and the 31st Mersenne prime, $2^{216,091} - 1$ (65,050 digits), also discovered by Slowinski, in 1985.

The 32nd known is $2^{756,839} - 1$, (227,832 digits) discovered by Slowinski and Gage in 1992. [JRM v25 262]

$2^{1,257,787} - 1$ [378,632 digits]

The second largest known prime number, and the 34th known Mersenne prime, announced in September 1996. It was found, like several previous records, by David Slowinski and Paul Gage, at Cray Research in Wisconsin, USA.

$2^{1,398,269} - 1$ [420,921 digits]

The highest known prime number, and the 35th known Mersenne prime, announced on 13 November 1996 in Paris by Joel Armengaud, who discovered it with his colleague George Woltman and many others, working as a team via the Internet.

9^{9^9}   [369,693,100 digits]

The largest number in decimal notation that can be represented without using more than 3 digits, with no additional symbols.

C. A. Laisant showed in 1906 that this number has 369,693,100 digits. In 1947 H. S. Uhler calculated and published the value of log 9^{9^9} to 250 decimal places.

Horace Scuder Uhler, Professor of Physics at Yale University, devoted much of his spare time to calculating an extraordinary variety of mathematical numbers, such as logarithms, reciprocals, roots, to immense numbers of decimal places. He found it relaxing. He found the calculation of log 9^{9^9} doubly relaxing – he did it in between testing for factors of Mersenne numbers such as M_{157}, which he showed to be composite. [*Mathematics Teacher*, April 1953]

1^{billion}

A gigaplex is equal to 1 with a billion zeros.

F_{23471}

The largest known composite Fermat number, discovered by Keller in 1984. It has the factor $5 \times 2^{23,473} + 1$, and more than 10^{7000} digits. [Ribenboim 73]

$10^{10^{10^{34}}}$

Skewes' number

The number of primes less than or equal to n is approximately $\int_0^n \dfrac{dx}{\log x}$.

For small values of n, into the tens of millions, this approximation is an overestimate, but this is not always so. J. E. Littlewood proved in 1914 his famous theorem that it switches from being an overestimate to an underestimate and back again an infinite number of times, if, of course, you go high enough. How high? Skewes proved in 1933 that the first switch occurs before n reaches $10^{10^{10^{34}}}$, though he had to assume the truth of the famous Riemann hypothesis.

At the time this was an extraordinarily large number. Hardy thought it 'the largest number which has ever served any definite purpose in mathematics', and suggested that if a game of chess was played with all the particles in the universe as pieces, one move being the interchange of a pair of particles, and the game terminating when the same position recurred for the 3rd time, the number of possible games would be about Skewes' number.

By way of comparison, the number of particles in the universe has been estimated in recent years as a trifling 10^{80} to 10^{87}.

Skewes' number is dwarfed by many numbers now appearing in problems in combinatorics. [Boas, in Honsberger, *Mathematical Plums*, Mathematical Association of America, 1979]

Skewes' number is also out of date as an upper limit. In 1986 te Riele showed that between $6\cdot62 \times 10^{370}$ and $6\cdot69 \times 10^{370}$ there are more than 10^{180} consecutive integers for which it is an underestimate. [Ribenboim 180]

$3 \uparrow \uparrow \uparrow 3$ etc., etc.
Graham's number

The World Champion largest number, listed in the latest *Guinness Book of Records*, is an upper bound, derived by R. L. Graham, from a problem in a part of combinatorics called Ramsey theory.

Graham's number cannot be expressed using the conventional notation of powers, and powers of powers. If all the material in the universe were turned into pen and ink it would not be enough to write the number down. Consequently, this special notation, devised by Donald Knuth, is necessary.

$3 \uparrow 3$ means '3 cubed', as it often does in computer printouts.

$3 \uparrow \uparrow 3$ means $3 \uparrow (3 \uparrow 3)$, or $3 \uparrow 27$, which is already quite large: $3 \uparrow 27 = 7,625,597,484,987$, but is still easily written, especially as a tower of 3 numbers: 3^{3^3}.

$3 \uparrow \uparrow \uparrow 3 = 3 \uparrow \uparrow (3 \uparrow \uparrow 3)$, however, is $3 \uparrow \uparrow 7,625,597,484,987 = 3 \uparrow 3 \uparrow 3 \uparrow 3 \uparrow \ldots 7,625,597,484,987$ times.

$3 \uparrow \uparrow \uparrow \uparrow 3 = 3 \uparrow \uparrow \uparrow (3 \uparrow \uparrow \uparrow 3)$, of course. Even the tower of exponents is now unimaginably large in our usual notation, but Graham's number only starts here.

Consider the number $3 \uparrow \uparrow \uparrow \ldots \uparrow \uparrow \uparrow 3$ in which there are $3 \uparrow \uparrow \uparrow \uparrow 3$ arrows. A largish number!

Next construct the number $3 \uparrow \uparrow \uparrow \ldots \uparrow \uparrow \uparrow 3$ where the number of arrows is the previous $3 \uparrow \uparrow \uparrow \ldots \uparrow \uparrow \uparrow 3$ number.

An incredible, ungraspable number! Yet we are only 2 steps away from the original ginormous $3 \uparrow \uparrow \uparrow \uparrow 3$. Now continue this process, making the number of arrows in $3 \uparrow \uparrow \uparrow \ldots \uparrow \uparrow \uparrow 3$ equal to the number at the previous step, until you are 63 steps, yes, *sixty-three*, steps from $3 \uparrow \uparrow \uparrow \uparrow 3$. That is Graham's number.

There is a twist in the tail of this true fairy story. Remember that Graham's number is an upper bound, just like Skewes' number. What is likely to be the actual answer to Graham's problem? Gardner quotes the opinions of the experts in Ramsey theory, who suspect that the answer is – amazingly – 6.

['Mathematical Games', *Scientific American*, Nov. 1977]

1 The First 100 Triangular Numbers, Squares and Cubes

	triangular	square	cube
1	1	1	1
2	3	4	8
3	6	9	27
4	10	16	64
5	15	25	125
6	21	36	216
7	28	49	343
8	36	64	512
9	45	81	729
10	55	100	1000
11	66	121	1331
12	78	144	1728
13	91	169	2197
14	105	196	2744
15	120	225	3375
16	136	256	4096
17	153	289	4913
18	171	324	5832
19	190	361	6859
20	210	400	8000
21	231	441	9261
22	253	484	10648
23	276	529	12167
24	300	576	13824
25	325	625	15625
26	351	676	17576
27	378	729	19683
28	406	784	21952
29	435	841	24389
30	465	900	27000
31	496	961	29791
32	528	1024	32768
33	561	1089	35937
34	595	1156	39304
35	630	1225	42875
36	666	1296	46656
37	703	1369	50653
38	741	1444	54872
39	780	1521	59319
40	820	1600	64000

	triangular	*square*	*cube*
41	861	1681	68921
42	903	1764	74088
43	946	1849	79507
44	990	1936	85184
45	1035	2025	91125
46	1081	2116	97336
47	1128	2209	103823
48	1176	2304	110592
49	1225	2401	117649
50	1275	2500	125000
51	1326	2601	132651
52	1378	2704	140608
53	1431	2809	148877
54	1485	2916	157464
55	1540	3025	166375
56	1596	3136	175616
57	1653	3249	185193
58	1711	3364	195112
59	1770	3481	205379
60	1830	3600	216000
61	1891	3721	226981
62	1953	3844	238328
63	2016	3969	250047
64	2080	4096	262144
65	2145	4225	274625
66	2211	4356	287496
67	2278	4489	300763
68	2346	4624	314432
69	2415	4761	328509
70	2485	4900	343000
71	2556	5041	357911
72	2628	5184	373248
73	2701	5329	389017
74	2775	5476	405224
75	2850	5625	421875
76	2926	5776	438976
77	3003	5929	456533
78	3081	6084	474552
79	3160	6241	493039
80	3240	6400	512000
81	3321	6561	531441
82	3403	6724	551368
83	3486	6889	571787
84	3570	7056	592704

	triangular	*square*	*cube*
85	3655	7225	614125
86	3741	7396	636056
87	3828	7569	658503
88	3916	7744	681472
89	4005	7921	704969
90	4095	8100	729000
91	4186	8281	753571
92	4278	8464	778688
93	4371	8649	804357
94	4465	8836	830584
95	4560	9025	857375
96	4656	9216	884736
97	4753	9409	912673
98	4851	9604	941192
99	4950	9801	970299
100	5050	10000	1000000

2 The First 20 Pentagonal, Hexagonal, Heptagonal and Octagonal Numbers

	pentagonal	*hexagonal*	*heptagonal*	*octagonal*
1	1	1	1	1
2	5	6	7	8
3	12	15	18	21
4	22	28	34	40
5	35	45	55	65
6	51	66	81	96
7	70	91	112	133
8	92	120	148	176
9	117	153	189	225
10	145	190	235	280
11	176	231	286	341
12	210	276	342	408
13	247	325	403	481
14	287	378	469	560
15	330	435	540	645
16	376	496	616	736
17	425	561	697	833
18	477	630	783	936
19	532	703	874	1045
20	590	780	970	1160

3 The First 40 Fibonacci Numbers

	Fibonacci
1	1
2	1
3	2
4	3
5	5
6	8
7	13
8	21
9	34
10	55
11	89
12	144
13	233
14	377
15	610
16	987
17	1597
18	2584
19	4181
20	6765
21	10946
22	17711
23	28657
24	46368
25	75025
26	121393
27	196418
28	317811
29	514229
30	832040
31	1346269
32	2178309
33	3524578
34	5702887
35	9227465
36	14930352
37	24157817
38	39088169
39	63245986
40	102334155

4 The Prime Numbers less than 1000

2	3	5	7	11	13	17	19	23	29	31	37
41	43	47	53	59	61	67	71	73	79	83	89
97	101	103	107	109	113	127	131	137	139	149	151
157	163	167	173	179	181	191	193	197	199	211	223
227	229	233	239	241	251	257	263	269	271	277	281
283	293	307	311	313	317	331	337	347	349	353	359
367	373	379	383	389	397	401	409	419	421	431	433
439	443	449	457	461	463	467	479	487	491	499	503
509	521	523	541	547	557	563	569	571	577	587	593
599	601	607	613	617	619	631	641	643	647	653	659
661	673	677	683	691	701	709	719	727	733	739	743
751	757	761	769	773	787	797	809	811	821	823	827
829	839	853	857	859	863	877	881	883	887	907	911
919	929	937	941	947	953	967	971	977	983	991	997

5 The Factorials of the Numbers 1 to 20

0!	1
1!	1
2!	2
3!	6
4!	24
5!	120
6!	720
7!	5,040
8!	40,320
9!	362,880
10!	3,628,800
11!	39,916,800
12!	479,001,600
13!	6,227,020,800
14!	87,178,291,200
15!	1,307,674,368,000
16!	20,922,789,888,000
17!	355,687,428,096,000
18!	6,402,373,705,728,000
19!	121,645,100,408,832,000
20!	2,432,902,008,176,640,000

6 The Decimal Reciprocals of the Primes from 7 to 97

1/7 = ·1̇42857̇
1/11 = ·0̇9̇
1/13 = ·0̇7692̇3̇
1/17 = ·0̇588235294117647̇
1/19 = ·0̇52631578947368421̇
1/23 = ·0̇434782608695652173913̇
1/29 = ·0̇344827586206896551724137931̇
1/3 1= ·0̇32258064516129̇
1/37 = ·0̇27̇
1/41 = ·0̇2439̇
1/43 = ·0̇23255813953488372093̇
1/47 = ·0̇212765957446808510638297872340425531914893617̇
1/53 = ·0̇188679245283̇
1/59 = ·0̇1694915254237288135593220338983050847457627118644067796̇6̇
1/61 = ·0̇163934426229508196721311475409836065573770491803278688524̇5̇9̇
1/67 = ·0̇149253731343283582089552238805597̇
1/71 = ·0̇14084507042253521126760563380281̇6̇9̇
1/73 = ·0̇1369863̇
1/79 = ·0̇12658227848̇1̇
1/83 = ·0̇120481927710843373493975903614457831325̇3̇
1/89 = ·0̇112359550561797752808988764044943820224719̇1̇
1/97 = ·0̇10309278350515463917525773195876288659793814432989690721649̇4̇
          8453608247422680412371134020618556̇7̇

7 The Factors of the Repunits from 11 to R$_{40}$

2	11
3	3.37
4	11.101
5	41.271
6	3.7.11.13.37
7	239.4649
8	11.73.101.137
9	3.3.37.333667
10	11.41.271.9091
11	21649.513239
12	3.7.11.13.37.101.9901
13	53.79.265371653
14	11.239.4649.909091
15	3.31.37.41.271.2906161
16	11.17.73.101.137.5882353
17	2071723.5363222357
18	3.3.7.11.13.19.37.52579.333667
19	1111111111111111111
20	11.41.101.271.3541.9091.27961
21	3.37.43.239.1933.4649.10838689
22	11.11.23.4093.8779.21649.513239
23	11111111111111111111111
24	3.7.11.13.37.73.101.137.9901.99990001
25	41.271.21401.25601.182521213001
26	11.53.79.859.265371653.1058313049
27	3.3.3.37.757.333667.440334654777631
28	11.29.101.239.281.4649.909091.121499449
29	3191.16763.43037.62003.77843839397
30	3.7.11.13.31.37.41.211.241.271.2161.9091.2906161
31	2791.6943319.57336415063790604359
32	11.17.73.101.137.353.449.641.1409.69857.5882353
33	3.37.67.21649.513239.1344628210313298373
34	11.103.4013.2071723.5363222357.21993833369
35	41.71.239.271.4649.123551.102598800232111471
36	3.3.7.11.13.19.37.101.9901.52579.333667.999999000001
37	2028119.247629013.2212394296770203368013
38	11.909090909090909091.11111111111111111111
39	3.37.53.79.265371653.900900900900990990990991
40	11.41.73.101.137.271.3541.9091.27961.1676321.5964848081

Reprinted from *Contemporary Mathematics*, 1983, vol. 12, 'Factorizations of $b^n \pm 1$ $b = 2, 3, 5,$ 6, 7, 10, 11, 12 up to High Powers', John Brillhart, D. H. Lehmer, John L. Selfridge, Bryant Tuckerman, and S. S. Wagstaff, Jr, by permission of the American Mathematical Society.

8 The Proper Factors, where Composite, and the Values of the Functions $\phi(n)$, $d(n)$ and $\sigma(n)$

n	Factors	$\phi(n)$	$d(n)$	$\sigma(n)$
1	—	1	1	1
2	—	1	2	3
3	—	2	2	4
4	2^2	2	3	7
5	—	4	2	6
6	2.3	2	4	12
7	—	6	2	8
8	2^3	4	4	15
9	3^2	6	3	13
10	2.5	4	4	18
11	—	10	2	12
12	$2^2.3$	4	6	28
13	—	12	2	14
14	2.7	6	4	24
15	3.5	8	4	24
16	2^4	8	5	31
17	—	16	2	18
18	2.3^2	6	6	39
19	—	18	2	20
20	$2^2.5$	8	6	42
21	3.7	12	4	32
22	2.11	10	4	36
23	—	22	2	24
24	$2^3.3$	8	8	60
25	5^2	20	3	31
26	2.13	12	4	42
27	3^3	18	4	40
28	$2^2.7$	12	6	56
29	—	28	2	30
30	2.3.5	8	8	72
31	—	30	2	32
32	2^5	16	6	63
33	3.11	20	4	48
34	2.17	16	4	54
35	5.7	24	4	48
36	$2^2.3^2$	12	9	91
37	—	36	2	38
38	2.19	18	4	60
39	3.13	24	4	56
40	$2^3.5$	16	8	90

n	Factors	$\phi(n)$	$d(n)$	$\sigma(n)$
41	—	40	2	42
42	2.3.7	12	8	96
43	—	42	2	44
44	$2^2.11$	20	6	84
45	$3^2.5$	24	6	78
46	2.23	22	4	72
47	—	46	2	48
48	$2^4.3$	16	10	124
49	7^2	42	3	57
50	2.5^2	20	6	93
51	3.17	32	4	72
52	$2^2.13$	24	6	98
53	—	52	2	54
54	2.3^2	18	8	120
55	5.11	40	4	72
56	$2^3.7$	24	8	120
57	3.19	36	4	80
58	2.29	28	4	90
59	—	58	2	60
60	$2^2.3.5$	16	12	168
61	—	60	2	62
62	2.31	30	4	96
63	$3^2.7$	36	6	104
64	2^6	32	7	127
65	5.13	48	4	84
66	2.3.11	20	8	144
67	—	66	2	68
68	$2^2.17$	32	6	126
69	3.23	44	4	96
70	2.5.7	24	8	144
71	—	70	2	72
72	$2^3.3^2$	24	12	195
73	—	72	2	74
74	2.37	36	4	114
75	3.5^2	40	6	124
76	$2^2.19$	36	6	140
77	7.11	60	4	96
78	2.3.13	24	8	168
79	—	78	2	80
80	$2^4.5$	32	10	186
81	3^4	54	5	121
82	2.41	40	4	126
83	—	82	2	84
84	$2^2.3.7$	24	12	224

n	Factors	$\phi(n)$	$d(n)$	$\sigma(n)$
85	5.17	64	4	108
86	2.43	42	4	132
87	3.29	56	4	120
88	$2^3.11$	40	8	180
89	—	88	2	90
90	$2.3^2.5$	24	12	234
91	7.13	72	4	112
92	$2^2.23$	44	6	168
93	3.31	60	4	128
94	2.47	46	4	144
95	5.19	72	4	120
96	$2^5.3$	32	12	252
97	—	96	2	98
98	2.7^2	42	6	171
99	$3^2.11$	60	6	156
100	$2^2.5^2$	40	9	217

Index

Only 3 decimal places are given in this index. **Bold** type indicates that the index term is defined under that entry.

Ramanujan   97·409

rational numbers   *see* fractions

reciprocals   2/3; 2; 35; 2520; *see also* fractions, Egyptian; harmonic series

reciprocals, decimal periods of   5; 7; 13; 17; 19; 23; 27; 31; 37; 49; 53; 81; 89; 97; 98; 99; 103; 729; 999; 1089; 47,619; 076,923; 142,857; 0588 . . . [15 digits]; 0526 . . . [17 digits]; 111 . . . [19 digits]; 0434 . . . [21 digits]

reciprocals, sums of   23·103; 77; 105; 272,400,600

rectangles, dissected   7; 21

repfigit   14; 129,572,008; 754,788,753,590,897

repunits   **11**; 111; 297; 4,937,775; 111 . . . [19 digits]; 111 . . . [23 digits]; 111 . . . [317 digits]; 111 . . . [1031 digits]

Rhind papyrus   2/3; 1·618; 2; 3·141; 7

Riemann hypothesis   **0·5**; 2·665; 5; 14·134; $10^{10^{10^{34}}}$

rigid framework   23

Roman numerals   0; 5; 10; 12; 50; 100; 500; 666; 2,300,000

Rubik's Cube   432 . . . [20 digits]

Russian peasant multiplication   2

St Ives   7

Sandreckoner, The 10^{51}

self-descriptive numbers   10,213,223; 1,031,223,314

semi-perfect numbers   12; **20**; 104; 945

semi-primes   **33**; 818

sequence, recursive   43

shuffles   7

Sieve of Eratosthenes   33

Skewes' number $10^{10^{10^{34}}}$

Smith numbers   728; 4,937,775; 12,345,554,321

sociable numbers   28; 276; **12,496**; 14,316

Sophie Germain primes   89; 554,688,278,429; 191 . . . [39 digits]; 390 . . . [1811 digits]

space group   219

spheres, packing of   0·740; 12; 24; 196,560

spheres, volume of   5; 5·256

square-free   0·607; 29

square pyramidal numbers   14; 24; **55**; 91; 4900; 208,335; 399,877,410,625

square roots   0·123; 1·414; 1·732; 10; 30739; *see also* squares

squares   4; 9; 10; 11; 12; 15; 16; 17; 25; 36; 45; 49; 81; 118; 120; 121; 169; 196; 481; 1127; 1225; 1444; 1681; 2025; 4523; 7744; 8281; 12,321; 19,600; 51,984; 74,162; 183,184; 621,770; 698,896; 1,048,576; 60,996,100; 1,026,753,849; 1,857,437,604; 6,661,661,161; 444,171,597,444; 260 . . . [20 digits]

squares, differences   10; 111

squares, geometrical   4; 112

squares, sums of   3; 4; 5; 7; 9; 10; 24; 25; 27; 33; 50; 55; 65; 85; 125; 128; 145; 188; 232; 250; 325; 365; 666; 720; 1105; 1331; 2601; 9010; 17,163; 103,823

Steiner's problem   1·444; 6·196

READ MORE IN PENGUIN

In every corner of the world, on every subject under the sun, Penguin represents quality and variety – the very best in publishing today.

For complete information about books available from Penguin – including Puffins, Penguin Classics and Arkana – and how to order them, write to us at the appropriate address below. Please note that for copyright reasons the selection of books varies from country to country.

In the United Kingdom: Please write to *Dept. EP, Penguin Books Ltd, Bath Road, Harmondsworth, West Drayton, Middlesex UB7 0DA*

In the United States: Please write to *Consumer Sales, Penguin USA, P.O. Box 999, Dept. 17109, Bergenfield, New Jersey 07621-0120.* VISA and MasterCard holders call 1-800-253-6476 to order Penguin titles

In Canada: Please write to *Penguin Books Canada Ltd, 10 Alcorn Avenue, Suite 300, Toronto, Ontario M4V 3B2*

In Australia: Please write to *Penguin Books Australia Ltd, P.O. Box 257, Ringwood, Victoria 3134*

In New Zealand: Please write to *Penguin Books (NZ) Ltd, Private Bag 102902, North Shore Mail Centre, Auckland 10*

In India: Please write to *Penguin Books India Pvt Ltd, 706 Eros Apartments, 56 Nehru Place, New Delhi 110 019*

In the Netherlands: Please write to *Penguin Books Netherlands bv, Postbus 3507, NL-1001 AH Amsterdam*

In Germany: Please write to *Penguin Books Deutschland GmbH, Metzlerstrasse 26, 60594 Frankfurt am Main*

In Spain: Please write to *Penguin Books S. A., Bravo Murillo 19, 1° B, 28015 Madrid*

In Italy: Please write to *Penguin Italia s.r.l., Via Felice Casati 20, I–20124 Milano*

In France: Please write to *Penguin France S. A., 17 rue Lejeune, F–31000 Toulouse*

In Japan: Please write to *Penguin Books Japan, Ishikiribashi Building, 2–5–4, Suido, Bunkyo-ku, Tokyo 112*

In South Africa: Please write to *Longman Penguin Southern Africa (Pty) Ltd, Private Bag X08, Bertsham 2013*

BY THE SAME AUTHOR

The Penguin Dictionary of Curious and Interesting Geometry

What do the Apollonian gasket, Dandelin spheres, interlocking polyominoes, Poncelet's porism, Fermat points, Fatou dust, the Voderberg tessellation, the Euler line and the unilluminable room have in common?

They all appear among the hundreds of shapes, figures, objects, theorems, patterns and properties in this collection of geometrical gems. From the simple circle to fiendish fractals, from billiard balls bouncing round a cube to geometry with matchsticks, from Pythagoras to Penrose tilings to pursuit curves, they are all here, with a comprehensive index to lead you to that triangle thingumajig, you know, the one where all the points lie on a line . . .

'A wonderful, strange experience. But be careful . . . The entries shoot out at you like psychedelic bullets. Each one undermines your confidence in your understanding of the world a little bit more. And if you fail to dodge some of the mindbending slugs, you may end up mindbroke . . . this collection of oddities is so odd it verges on greatness' – *New Scientist*

The Penguin Book of Curious and Interesting Puzzles

Wherever there are human beings, setting and solving problems have always been among their principal passions. The first surviving 'think of a number' problem dates back to an Egyptian papyrus of around 1650 BC . . .

This collection of logical and mathematical puzzles, none requiring specialist knowledge or more than pencil, paper and a few counters, brings together examples from the earliest times up to the present day. Some are truly demanding, others require a single 'easy' but easily missed lateral leap – so turn to the solutions or endure days of delicious frustration! Whether they concern Prisoner's Dilemmas, fast-breeding rabbits, liars and truthtellers or Prince Rupert's cube, one thing is sure: endless entertainment is guaranteed.

'It is a book of fatal attraction. Once hooked you will spend valuable time getting the wrong answer to the Bemusing Bolts problem . . . Look up the answers and astonish friends' – *New Scientist*

BY THE SAME AUTHOR

The Penguin Book of Curious and Interesting Mathematics

This fascinating compendium of strange facts and anecdotes includes probability paradoxes, jumbled Shakespearean sonnets, African river-crossing problems, monkeys and typewriters, googols and googolplexes, not to mention a definitive proof that 1 and 1 makes 2.

Spanning the centuries, David Wells introduces a collection of choice eccentrics who calmed their nerves with algebra or used sextants to measure the buttocks of Hottentot women. Along with Newton's views on chance and chaos and scenes from the life of Pythagoras, he presents maths in the Bible as well as maths and misogyny, madness and the military.

With Wells's unique gift for making mathematics lively and accessible, this book will provide both endless entertainment and a window on a weird and wonderful world.

You Are a Mathematician

Anyone familiar with numbers, circles, straight lines and squares can start becoming a mathematician. 'All you have to do,' claims Wells, 'is to learn to look at these objects with some insight and imagination, maybe do a few experiments, and be able to draw reasonable conclusions . . .'

This entertaining and informative introduction to mathematics begins with the secrets of triangles and the dazzling patterns formed by even the simplest numbers. It examines polyhedral cheeses, reverse Koch snowflakes and Rabbi Moses' box, takes readers on 'a journey from the Greek mathematicians to quantum theory', and concludes with a challenging adventure game. Scientists searching for the truth (or useful approximations) find mathematics an invaluable scientific tool. Yet mathematical thinking is also very much like a game, relying on cunning tactics, deep strategy and brilliant combinations as much as on observation, analogy and informed guesswork. This book is an ideal guide to its potential and pleasures.